Lecture Notes in Computer Science

Lecture Notes in Computer Science

Edited by G. Goos and J. Hartmanis

275

A. N. Habermann U. Montanari (Eds.)

System Development and Ada

CRAI Workshop on Software Factories and Ada
Capri, Italy, May 26–30, 1986
Proceedings

Springer-Verlag

Berlin Heidelberg New York London Paris Tokyo

Editors

A. Nico Habermann
Department of Computer Science, Carnegie Mellon University
Pittsburg, Pennsylvania 15213, USA

Ugo Montanari
Dipartimento di Informatica, Universitá di Pisa
Corso Italia 40, I-56100 Pisa, Italy

CR Subject Classification (1987): D.2.2, D.2.6, D.2.9, D.2.10

ISBN 3-540-18341-8 Springer-Verlag Berlin Heidelberg New York
ISBN 0-387-18341-8 Springer-Verlag New York Berlin Heidelberg

Library of Congress Cataloging-in-Publication Data. Workshop on Software Factories and Ada
(1986: Capri, Italy) System development and Ada. (Lecture notes in computer science; 275)
Includes bibliographies. 1. System design–Congresses. 2. Computer software–Development–
Congresses. 3. Ada (Computer program language)–Congresses. I. Habermann, A. Nico, 1932-.
II. Montanari, U. (Ugo) III. Title. IV. Series.
QA76.9.S88W66 1986 004.2'5 87-23397
ISBN 0-387-18341-8 (U.S.)

© Springer-Verlag Berlin Heidelberg 1987
Printed in Germany

Printing and binding: Druckhaus Beltz, Hemsbach/Bergstr.
2145/3140-543210

Preface

The collection of papers published in this book was initially presented at the workshop on Software Factories and Ada held on Capri in May 1986. The collection consists of three groups: the first five papers concern software development environments, the next four papers deal with formal methods for software development and the last three discuss activities related to the Ada language.

Although Ada plays an important role in the papers, it is clear that the work presented here goes far beyond the limits of a programming language. The environment papers show a variety of approaches, ranging from language-specific environments to multi-language and methodology-driven environments. All environments presented in the first five papers are operational and are commercially available.

The papers on formal methods show the great variety of ideas that exist to assure the accurate correspondence of specification and implementation. Of particular interest are the papers that do not just address the programming product, but develop formal methods for the design process as well. Interesting ideas are presented on planning the design process and on supporting project management by formal tools.

The impressive task of Ada compiler validation is described in the first of the last group of papers. The very last paper in this group presents an interesting approach to specifying and testing events in Ada tasks. The paper shows that this approach extends to languages other than Ada.

The authors want to express their gratitude to the Italian hosts, in particular CRAI (Consorzio per le Ricerche e le Applicazioni di Informatica), who made the local arrangements. We are also grateful for the cooperation of the Commission of the European Communities in Brussels, Belgium and the Software Engineering Institute in Pittsburgh, Pennsylvania.

The Editors,   A. Nico Habermann
                 Ugo Montanari

Table of Contents

Design of the Rational Environment

James E. Archer, Jr.[1]

1. Introduction

The development of very large software systems challenges human intellect and creativity. It has been described as one of the most complex activities undertaken by Man. Considerable research effort has been devoted to solving the problems involved in the construction of such large systems. Unfortunately, while much of the resulting technology is available in the literature, it is not widely used [16]. Reducing theory to practice is always difficult; the rate at which this has been accomplished for software seems particularly discouraging. These difficulties prompted the Department of Defense to start the STARS program [7] and to establish the Software Engineering Institute [3] to focus on improving the state of practice.

2. Motivation

In 1981, Rational set out to produce an interactive environment that would improve productivity for the development of large software systems. The mission was to create an environment that supported high-level language development of systems developed using modern software engineering principles. This mission was built on the belief in recent advances in programming languages, methods, and environments.

In designing the Rational Environment, object-oriented design [4], abstraction [9], information hiding [5], and reusability [10] were important both in terms of use in our design and as methods to be supported. Prototyping [11] was of particular importance because it gave access to the advantages of the environment and its component technologies, at the earliest possible time.

The language to be supported was Ada. That was an easy choice. Ada appeared to be the latest and best engineered language for building large systems [1]. Ada's separation of specification and implementation offered important advantages, including the ability to use the language from intial system design through implementation, and a realistic opportunity for reusability [8].

Experience with research programming environments had shown that access to a set of integrated facilities could greatly leverage the ability of individuals to produce systems. The most widely used of these environments supported interpretive or dynamically typed languages, most notably Lisp [13]. Research efforts to support more appropriate, strongly typed languages were interesting, but they centered mostly on interpretive implementations for student subsets [2, 14]. Even so, the benefits of these systems suggested the feasibility of building a compilation-based environment for team development of large systems.

---

[1]Rational, 1501 Salado Drive, Mountain View, California 94043

From the outset, it was evident that the Environment to be created was the same kind of system whose development it was intended ultimately to facilitate. Although the Environment would obviously not be available for use in its own development -- at least not during the early stages -- the technology themes, the Ada language, and and the general methodological guidelines could be applied.

As a result, the initial task was to construct a set of development tools to support Ada development in a conventional, batch-oriented manner. That was done. We don't think of the resulting tool set as an "environment"; however, it does constitute an APSE in the Stoneman sense [12]: it includes a validated compiler, and it is comparable to other commercially available Ada compilation systems. The development of this tool set involved more than 300,000 lines of Ada code; building it helped to deepen our understanding of the problems and opportunities associated with the evolution of the Rational Environment.

3. Environment Characteristics

The Rational Environment is the operating software for the Rational R1000, a time-shared computer implementing a proprietary, Ada-oriented architecture. It is written entirely in Ada, with considerably less than 1% of its statements being machine code insertions.

The Environment is the system interface; all users of the system use the same facilities. Although general-purpose computing is well supported, the system is designed to be used by project-related personnel with some interest in and facility with Ada and programming language concepts.

3.1. Ada Framework

The Environment directory structure is hierarchically organized. Names in this structure are Ada simple identifiers separated by periods, in the style of Ada qualified names. Directories may contain objects of a variety of types. One common object type is *File*; another is *Ada*. The characteristics of Ada objects and their storage are central to the nature of the Environment.

The names of Ada objects are the same as the names of the Ada units that they represent. Ada objects corresponding to library units have two parts: visible part and body. Separate subunits of Ada units are children of their parent Ada unit and are named as such. The same name is used to refer to the unit when it is edited, compiled, or executed. All of the units residing in the same directory substructure constitute an Ada library, and there are provisions for creating libraries, hierarchical or otherwise, from multiple simple libraries.

Program objects are stored in the Environment as attributed DIANA tree [6]. That fact, in conjunction with the support of an object-oriented view of the unit throughout its life, provides ready access to the program structure in ways that make it possible to create some rather powerful facilities. These include Ada-specific editing operations, incremental compilation, compilation ordering and interconnection facilities, and direct execution of Ada statements.

3.2. Compilation

In the traditional compilation model, files containing the program source are read into a sequence of software tools that produce various processed forms of the original program. During this process, new objects with new names are created and the system developer must be able to track the correspondence between the current program text and the current executable version. In contrast, the Environment compilation model centers on Ada units as definable objects that are transformed by editing and compilation between three principal states: *source*, *installed*, and *coded*.[2]

A *source unit* is one that has been parsed, but has yet to be compiled. It isn't just another file: it's a DIANA tree sufficient to support interactive syntax checking and operations that depend on the logical structure of the unit. Having the Environment maintain this structure makes it convenient to keep units syntactically consistent[3], greatly reducing the time lost trying to compile units that contain syntax errors.

A unit in the *installed* state has passed all of the semantic checks required by the Ada language. Ada was explicitly designed to support static inter-module consistency checking of separate units. To be installed, the Ada unit must have passed all of the applicable semantic checks with regard to other units that it references. Getting units semantically consistent -- and keeping them that way -- is one of the major programming activities in the development of Ada-based systems. Assuring this consistency is a principal intellectual activity of the developer and a signficant consumer of machine compilation resources. One of the signal strengths of the Rational Environment is its ability to provide assistance in reducing both the human and machine resources required to understand and maintain semantic consistency.

Coded units are ready for execution. Once a unit is installed, coding is just a matter of computational time; no further intellectual effort is involved. The separate coded state is used to save computational resources. During development, it is common for a unit to be changed and referenced many times between executions. Coding every time semantic checking is done wastes resources and, consequently, reduces interactivity. For changes that are to executed directly, it is always possible to request direct transformation from the source state to coded.

The Environment supports a spectrum of compilation paradigms:

- Batch installation and coding with fully automatic ordering
- Editor-based installation and coding of individual units

---

[2]Compilation for targets other than the R1000 may involve more than these three states.

[3]The editor allows arbitrary text transformations of Ada units to allow transformations whose intermediate states are not syntactically correct. Ada units can also be saved with syntax errors.

- Incremental, statement/declaration-level changes to installed units

Implementation of each of these paradigms makes use of the Environment's knowledge of the structure of the units being processed to minimize effort while assuring complete correctness.

Independent of the language involved, the ability to make incremental changes to compiled units has an obvious and intuitive appeal. The effort to compile a change should be proportional to the size of the change. This is particularly important for Ada. In addition to conserving computational resources, this allows the Environment to provide immediate feedback in a way the assists the developer in the important process of maintaining the declarative structure of the program.

Another benefit to be derived from incremental operations is the ability to add new functionality to a specification with minimal compilation effort. An Ada system consists of a number of modules that are logically interconnected and must be kept semantically consistent. Adding a declaration to a package can have ramifications on large parts of the system. In text-based, batch compilation systems, Ada consistency requirements can result in the recompilation of all units that reference any of the declarations in the same package as the one being added. The Environment uses its knowledge of the Ada structure to significantly reduce the set of affected units without compromising correctness. Though the difference is typically less dramatic, there are examples of changes that can be done in seconds that would require hours of recompilation using less sophisticated technology. Providing this facility was one of the more interesting technical challenges in building the Environment [15], but it was certainly worth the effort.

Immediate semantic information about programs under development is not limited to the compilation process. Part of providing incremental semantics was building a database of declaration-level dependency information sufficient, in conjunction with the DIANA trees, to determine the legality and impact of incremental changes. This information turns out to be generally useful. For installed Ada units, the relationships between declarations and their uses is a matter of great interest. Given the rich structure of Ada naming (use clauses, renaming declarations, overloading), it isn't possible, much less desirable, to keep track of these relationships on the basis of the program text. Using the compilation dependency information, these relationships can be checked interactively. Two operations are of particular interest: *Definition* and *Show_Usage*.

Definition is the name of an operation to show the declarations of an object that is referenced somewhere in an Ada unit. As typically used, the user points to the reference of interest[4] and presses the key to provide its definition. The declaration of the object referenced is brought onto the screen in the context of its Ada unit. It is also possible to find the implementation of the declaration. Definition is very useful in refreshing

---

[4]It is also possible to type its name if there is no immediate occurrence to point to.

familiar code in the user's mind; it is invaluable in developing an understanding of unfamiliar code. The Environments basic command for visiting objects of any type, for traversing the directory structure, or for changing contexts is a generalization of the Definition operation.

Show_Usage is the name of the operation that is in a sense the inverse of Definition: it provides the set of program locations that reference a particular declaration. This is a particularly useful interactive form of cross-reference. If only one unit references the declaration, the referencing unit is brought onto the screen with each of the references marked by underlining. This is the same mechanism that is used for reporting errors and there are convenient commands for visiting each of the marked sites in a unit. Where multiple units reference the declaration, a menu of units is present and the definition operation applied to any of the menu entries brings up a marked image of the indicated unit.

Show_Usage runs in time proportional to the number of first-level references, typically a second or two. It is an invaluable aid in determining the impact of an anticipated change.

3.3. Ada Command Language

Conventional interactive systems usually provide some sort of command shell through which specially prepared (and loaded) procedures can be invoked[5]. These procedures must either live in a simplified world without parameter passing or understand how to read arguments from the command line. In the former case, functionality is reduced; in the latter case, one such program calling another must understand how to invoke the shell and construct the appropriate parameter list to look as if it were entered to the shell. Further, parameter types are typically limited to some form of string.

In the Rational Environment, any coded visible subprogram can be executed simply by calling it, provided that the closure of units required by Ada rules is also coded. This feature has a profound effect on the accessibility of code for execution and testing. By unifying the shell/program interface to use the normal Ada parameter mechanisms, the interface is made both simpler and more powerful.

One salient benefit of this design is the consequential ability to use the richness of Ada semantics. The ability to reference the declarations in Ada units is not limited to procedures and functions; it extends to all Ada declarations: types, objects, constants, generics. The advantages that Ada has for the expression of application designs are available for the specification of the system interface or user-written utilities. This generality has far-reaching implications on the appearance, usage, and implementation of the system. References to procedures with complicated parameter profiles can be expressed using the name notation, parameter defaulting, and overloading. Interfaces to predefined packages, e.g., Text_IO, are just as easily invoked as commands to create files, using the same interface.

---

[5]Menu-based systems have similar, often more severe, limitations in this regard.

The full power of Ada is important to making the command interface work. An Ada-like interface that is limited to normal command-style entries might seem to be a good tradeoff between required implementation effort and the utility of the resulting interface. Closer inspection reveals the limitations of that strategy. Cutting isolated features from any language is a treacherous undertaking. As a simple example, private types are useful for providing abstract interfaces to system functionality, and having private types in the command interface turns out to be quite useful. This usefulness, in turn, depends on the ability to provide function results as parameters and, in many cases, to make them default reasonably. It is important to retain Ada's full power in the command interface -- another instance of the value of maintaining a *seamless web* when possible.

The command interface is an Ada declare block into which the user typically enters a single procedure call that is executed. In the general case, it is possible to write complete Ada programs using tasks, generics, or any other Ada construct in this block. The block is then compiled, and code is generated and executed.[6] The completeness of the facility is often exploited in learning Ada and determining *what would happen if ...*. Since the interface is in fact Ada, it is strongly typed: many common errors are detected during compilation rather than during execution. Facilities for syntactic and semantic completion are also provided.

3.4. Editor-Based User Interface

All interactions with the Rational Environment are through a general, object-oriented, multi-window editor. At one level, the editor provides familiar *what you see is what you get* on the images corresponding to the objects being edited or viewed. The text of the images can be modified directly using character, word, and line operations; portions of images can be copied or moved to other locations in the same or different images; there is a general search/replace interface. All of these capabilities allow the user to view and modify objects in a human-readable, textual form.

Many of the various types of objects in the system, most notably Ada, are stored using more interesting data structures than text. To support the transition from text editing to the underlying object representation, the editor supports an incremental change, parse, pretty print cycle. Changes to the text are saved for processing by type-specific editors that understand the syntax of the particular object. The changes are processed by the incremental parser to create consistent object structures. As necessary, the revised data structures are reflected onto the image with any corrections or embellishments that are deemed appropriate by the editor for the type. A type-specific editor, called an *object editor*, is available for each of the main object types. All of these implement similar editing cycles, but the operations, grammar, and semantics for Ada, discussed below, are the most interesting.

---

[6]There is a special fast path provided for a common subset of known procedures for which no code is generated. This covers about 80% of the command executions. Users are typically unaware of which path a particular command takes.

The actual operations provided for editing an object are logically separated into three classes: *image operations*, *type-sensitive operations*, and *type-specific operations*.

Image operations are the outer-level, character-oriented operations; these are the same for all object types.

Type-sensitive operations are those that are expected to be available for all object types, but depend on the characteristics of the type; these include edit, structural selection, detail control, and various state transformations.

Type-specific operations are provided by some types of objects where the characteristics of the type require additional functionality. Creating an object is type-specific.

A simple, commonly-used object editor is the one provided for the subclass of files corresponding to text. Its use with files, whether created by the editor or written by programs with Text_IO, is fairly conventional. As with all object editors for which direct text changes are supported, it allows the user to select text from other object types. This is frequently used for to include examples in documents, mail, or bug reports.

The text object editor is also responsible for dealing with Standard_Input and Standard_Output for interactively executed programs. In this mode, the user has full access to the features of the editor in providing input to programs and checking their output. Output is only discarded by user request, so the output from executing programs can be inspected long after execution completes.

One of the features of the editor interface is that it doesn't impose any particular interaction sequence on its users. As a result, it is possible to freely switch between objects being edited and executing programs. The input required by an executing program can be provided by copying the text from another object or from a previous run of the same program. To support multiple concurrent activities, all visible windows are kept current with the values of the underlying objects, including (optionally) scrolling windows into which program output is being generated. This makes it convenient, for instance, to maintain a window on a long-running command to monitor its progress while continuing to get work done on something else.

3.5. Ada Editing

Ada was designed to allow the specification and construction of complex systems that could be read, understood, and maintained. A person has to write the programs, preferably using the expressive capabilities that will serve well throughout the life of the code. The purpose of the Ada object editor is to make the writing as easy as possible.

By understanding the syntax of Ada, the editor is able to provide interactive syntax checking and completion. Syntactic completion is based on the notion that many tokens in the syntax are redundant; providing the additional tokens is only marginally harder than detecting their absence. For instance, most of the structures of Ada syntax are signaled by keywords or punctuation that bracket constructs; e.g., the existence of the keyword *if* implies the future existence of *end if* and at least one statement in between. The editor uses this information to provide the keyword structure and, if required, prompts for the expression and statement portions of the statement.

The resulting facility is logically very similar to operations provided by syntax-directed editors, but is stylistically similar to normal text editing and conveys the "feel" of text editing to the user. Syntactic correctness is only enforced at user request, but such requests are part of normal interactive editing. Used frequently, the program can be kept syntactically correct; when necessary, wholesale editing can take place without incurring checking overhead until the changes are believed to be complete. Prompts are presented in a special font and obligingly disappear when typed over, providing convenient reminders of code still to be written. Any attempt to execute a prompt raises an exception.

Although required less frequently, a powerful form of syntactic completion is provided to construct skeletal bodies from specifications. Completion saves typing the same procedure headers in both the visible part and the body. A related operation creates a private part with prompted completions for each of the private types in the package.

The logical extension of syntactic completion is *semantic completion*. Semantic completion fills out the contents of expressions, most commonly subprogram calls or aggregates, in a manner analogous to the way syntactic completion fills in the structural parts of the language. When making an incremental change in an installed or coded unit, it is possible to enter part of an expression, typically a procedure or function call, and request that the system fill in the parameter profile with prompts for parameters without defaults. In doing so, the system will provide the full name-notation presentation of the call, supporting good stylistic use of the language without requiring the user to do the additional typing.

3.6. Debugging

The Rational Environment supports debugging in the same spirit as the other parts of the programming process. Debugging a program is just like running it without the debugger, except that a different "execute" key is used. No special preparations are required to set programs up to be debugged. Debugging is not intrusive: two people can be debugging the same program at the same time without getting in each other's way. Interaction with the debugger is at the source level. Program locations are displayed by bringing up the Ada image of the statement and highlighting it. Variables and parameters can be displayed by selecting them and pressing the "Put" key or by entering a command with the name of the desired variable. The value displayed is presented as it would appear in program source: record values are printed as aggregates with field names; enumeration values are printed as the appropriate enumeration literal.

3.7. Host-Target Support

Although the R1000, which is the Environment's host, provides an attractive environment for the execution of Ada programs, the system was designed to support the development of programs that would also run on other targets. With the exception of the execution-time interface, the system provides all of the facilities described -- irrespective of the target.

To the user, editing and compilation appear the same for other targets as for the R1000. Indeed, the specification of the target of a compilation is a declarative property of the library and affects the content, but not the form, of the basic operations. A general compilation interface is provided that captures target dependencies in installation and coding sufficient to support the development of code generators for any desired target. The interface is encapsulated to allow the use of code generators written outside of Rational to be integrated with the Environment.

Execution and debugging for other targets are less easily specified, but the debugger architecture includes support for the same set of operations on targets connected by communication lines as for native R1000 programs. There is also provision for target-specific debugging operations in a manner analogous to that used by the editor to provide type-specific operations. A variant of this host-target strategy was used successfully in debugging the Environment in the early stages of its development.

3.8. Configuration Management and Version Control

Supporting an object-oriented view of Ada units implies support for configuration management and version control within the same integrated context. Previous experience with research environments suggested that programs need not be files, but all of those efforts were intended to be used by lone researchers working on prototype systems, not teams producing a product. Conventional systems solve the problem by separating program storage and version control from compilation; this separation is impractical without compromising the advantages the Environment provides for compilation, completion, and other development activities.

A distinct but related problem that arises in the development of a large system is control over the configuration to be compiled and executed. Early experience showed that the connectivity of a large Ada system (the environment itself) makes it attractive to break the system up into subsystems to allow changes in one part of the system to be tested before being used by another. Simply executed, this strategy provides some relief, but it still strains compilation resources at integration points. This strain was especially bothersome, since integration took place during a prototyping stage when long delays in reintegration were especially undesirable.

The solution that we devised to handle to this configuration problem was to break the system up into a *subsystems*. Each subsystem has a specification and implementation component in the same way that Ada packages do. The specification, called a *subsystem interface*, consists of a set of Ada visible parts that represented the exported functionality of the subsystem. The complete set of units corresponds to the body of the subsystem. As with Ada units, the contract made by the visible part must be fulfilled by the body, but the implementation of the body can be changed without recompilation of clients of the visible part. An extension to the notion of incremental change of visible parts within a subsystem is that of *upward-compatible changes*. An upward-compatible change is a new declaration that is added to a subsystem interface such that references compiled against a version of the interface without the new declaration will continue to work. New code can be begin to reference the additional functionality.

One very effective addition to the subsystem technology was the ability to hide the private parts of packages.[7] Private parts are instrumental in providing abstract interfaces whose underlying implementation can be changed without rewriting referencing code. This extension makes it possible to change the representation without recompiling the clients, just as if the completion of the type were in the body. For our code, this capability was particularly useful. It is common to have a package that exports private types whose completions are types exported from instantiation(s) of generics that are only referenced for this purpose. Closing the private part makes it unnecessary for the interface to appear to *with* the package exporting either the generics or the types involved. Reducing the *with* closure reduces the size of the interface while reinforcing the spirit of abstract interfaces.

This ability to compose a system of compatible subsystems that have not been directly compiled together greatly facilitates integration, especially since the assurance of semantic integrity is not lost. It does not directly address the version control problem, but leads to a version control policy based on a series of *views* -- configurations of the entire subsystem library, each spawned from the previous version of the view.

Experience with these mechanisms and experience with the compilation system have led to the construction of a more sophisticated form of view that combines the advantages of subsystems, reservation-model "source" management, and differential storage of changes to provide a facility that effectively combines the best of conventional version control with the advantages of subsystems for forming configurations. By managing views for the user, it is possible to provide support for these various forms of multiplicity in such a way that there seems to be more than one version only when differentiating configurations is part of the work at hand.

3.9. Life-Cycle Support and Extensibility

A general goal of the Rational Environment is to provide support all of the life-cycle activities involved in software development. The initial implementation effort has focused on support for detailed design through maintenance, taking care that the structures and facilities of the resulting environment were conducive to extension into other parts of the life cycle. Our experience has been that Ada, by itself, provides a strong and natural foundation for system design. It is especially useful to be able to compile the designs and trace through the dependencies.

---

[7]The R1000 architecture provides efficient support for this form of truly private type.

Many of the facilities that make the Environment attractive for programming also make it attractive for tool development and use. The access to DIANA and semantic information holds out the promise of building tools to analyze programs and and their development. The ability to construct interactive, editor-based interfaces has proved attractive and has helped in the process of providing useful interfaces and is likely to play a significant role in the provision of additional functionality related to full life-cycle support.

4. Experience

The Rational Environment itself consists of about 800,000 lines of Ada. Development of the Environment also required building about 700,000 lines of Ada to provide cross-development tools (compilers, debuggers) and hardware/microcode development tools (simulators, translators, analysis programs). The product was first shipped to customers in February of 1985. Several significant upgrades (involving greatly improved performance, increased functionality, and improved robustness) have been delivered since then.

This development has provided considerable experience in the use of Ada with modern software engineering practices. This experience can be summarized by the following statements:

1. Adoption of Ada and the software engineering practices referenced earlier has been somewhat more difficult than anticipated. Significant investment in tools, training, and experience has been required.

2. The benefits are very real. Improvements in productivity and quality have been evident in all phases of development: design, implementation, integration, test, and maintenance.

4.1. Early Ada Experience

In 1981 and early 1982, a series of programs was constructed: development and simulation tools and prototypes of high-risk components of the Environment. These programs typically consisted of 50-100K lines of Ada.

Ada proved to be an excellent language for applying the concepts of information hiding, data abstraction, and hierarchical decomposition based on levels of abstraction. The basic package mechanism, separation of specification and implementation, and private types allowed rapid construction and modification of large, modular programs.

Ada cannot force good design, but it does capture and clarify the decomposition and connectivity of programs, allowing early detection and correction of architectural flaws in the design. Ada became our primary design tool, particularly for detail design. With experience, we were able to produce high-quality designs quite rapidly.

The interaction between semantic checking and modularity has been the basis for continuing, significant improvements in productivity. Using modularity and type structure to capture design information increased somewhat the time required to first execute the program, but it also greatly increased the chances that the first execution would be productive. New arrivals to the project frequently complained that they weren't ever going to get their first program to compile, only to come back later amazed how quickly it worked after the first execution. When problems did arise at runtime, constraint checking allowed the errors to be detected early in execution. A common, effective debugging strategy was (and is) to run the program until an unexpected exception occurs; the problem is often evident with no additional information. Even when this is not the case, the modularity of most programs reduces uncertainty about interactions and allows much more rapid isolation of errors. It is also much easier to reason about the structure of programs and predict the consequences of a change.

Early experience also showed that all these wonderful benefits were not free. Ada semantic analysis is very expensive, increasing compilation times significantly relative to other languages. The early detection of interface and type-safety errors was handicapped by the use of batch compilation technology to report these errors. This confirmed our belief that an interactive environment for Ada with support for incremental compilation would greatly improve productivity.

4.2. Large-Scale Development and Integration

In 1983 and 1984, the development focus at Rational shifted from developing programs consisting of 10-100 packages to incrementally constructing and integrating a complete system made up of 30-40 subsystems, where each subsystem was the size of one of the earlier programs.

The system was decomposed hierarchically into five major layers, with each layer consisting of 5-8 subsystems. Although there were significant structural and interface changes over the life of the project, the basic architecture was surprisingly stable. This architecture allowed considerable parallelism in the overall development process and was instrumental in the evolution of our understanding of the configuration management and version control issues in developing large Ada systems.

At a very early point, the components of the system (or skeletons of the components) were integrated into a complete system. This initial system had very limited functionality, but allowed the basic architecture to be "debugged" before the entire system was constructed. This integration allowed system design issues such as storage management, concurrency, and error handling to be addressed very early in the development process. Early integration also served to stabilize major interfaces.

Development of the individual subsystems proceeded in parallel, with periodic integration to provide a new baseline for further development. The use of hierarchical decomposition allowed enough independence for development to proceed in parallel, while providing tight interface control to minimize integration problems. It was this integration process that led to the evolution of the subsystem concepts and supporting tools described in section 3.8.

Use of

1. the Ada language,
2. object-oriented design,
3. integrated configuration management and compilation, and
4. an incremental integration strategy

proved to be a very effective combination for achieving the end of this particular project.

4.3. Maintenance

The Rational Environment has been in field use for about 18 months in multiple releases. Supporting it has provided additional insight into the maintenance phase of a large Ada system. At Rational, maintenance was (and is) the responsibility of the original development team. It was crucial that new development proceed in parallel with maintenance without having to hire and train significant numbers of developers.

Our experience has indicated that Ada's greatest value may be in maintenance. In this particular case, *maintenance* included bug fixes and minor enhancements, addition of major new functionality, redesign and reimplementation of several subsystems to improve performance, and reorganization of parts of the user interface. Since initial product introduction, not only has it been possible to provide desired new functionality, but reliability and robustness have improved and overall system performance has been increased by more than a factor of 3.

Efforts to improve performance provided concrete examples of both the power and the associated dangers of modularity and abstraction. Bringing up a large system in minimum time was greatly facilitated by abstract interfaces and the ability to reuse code (Ada generics). There were several cases where performance-critical sections of code were operating through generics in multiple layers of the system. For each of these, a much faster implementation has been implemented without material change to the clients of the interface. Ironically, the same nominal strengths of modularity and abstraction that led to the problems contributed to their solution. Critical sections were completely redone and integrated into the system without major disruption. Abstraction is not an end in itself, but used carefully, it can help produce reliable, maintainable software that can also meet performance demanding performance requirements.

4.4. Experience Using the Rational Environment

Our experience using the Rational Environment has confirmed those advantages we foresaw when we started the project. Interactive syntactic and semantic information makes a tremendous difference in the ease of constructing programs and making changes to them. The ability to follow semantic references makes it easier to understand existing programs and the impact of changes. The integrated debugger makes it much easier to find bugs and test fixes quickly. Taken together, these facilities have helped greatly in reducing the impact of ongoing maintenance on our ability to produce new code. We anticipate similar improvements as we achieve the same level of integration and interactivity for configuration management and version control.

The Environment has also proved useful in introducing new personnel to the project and existing personnel to new parts of the system. New personnel benefit from the assistance with syntax and semantics; everyone benefits from the ability to traverse and understand the structure of unfamiliar software. It is often possible for someone completely unfamiliar with a body of code to use these facilities to understand it well enough to successfully diagnose and fix bugs in a matter of minutes.

Acknowledgments

The design and implementation of the Rational Environment was a group effort involving too many people to mention individually. Each member of the team contributed invaluable ideas, effort, and experience. It has been a challenging, rewarding, and enjoyable process.

Ada is a registered trademark of the U.S. Government (Ada Joint Program Office).

Rational is a registered trademark of Rational.

Rational Environment is a trademark of Rational.

References

1. *Reference Manual for the Ada Programming Language.* Washington, D.C., 1983. United States Department of Defense.

2. J.E. Archer, Jr. *The Design and Implementation of a Cooperative Program Development Environment.* Ph.D. Th., Cornell University, August 1981.

3. M. Barbacci, A. Habermann, and M. Shaw. "The Software Engineering Institute: Bridging Practice and Potential". *IEEE Software 2*, 6 (November 1985), 4-21.

4. O-J. Dahl and K. Nygaard. "SIMULA - An Algol-Based Simulation Language". *Comm. ACM 9*, 9 (September 1966).

5. D.L. Parnas. "On the Criteria to Be Used in Decomposing Systems into Modules". *Comm. ACM 15*, 3 (December 1972).

6. A. Evans, K. Butler, G. Goos, W. Wulf. DIANA Reference Manual. TL 83-4, Tartan Laboratories, Pittsburgh, Pa., 1983.

7. L. Druffel, S. Redwine, and W. Riddle. "The STARS Program: Overview and Rationale". *IEEE Computer 16*, 11 (November 1983), 21-29.

8. J. Ichbiah. "Rationale for the Design of the Ada Programming Language". *SIGPLAN Notices 14*, 6 (June 1979). Part B.

9. B. Liskov, and S. Zilles. "Specification Techniques for Data Abstractions". *IEEE Trans. on Software Eng. SE-1* (March 1975).

10. M. McIlroy. Mass-Produced Software Components. In *Software Engineering Concepts and Techniques*, NATO Conference on Software Engineering, 1969.

11. T. Standish and T. Taylor. Initial Thoughts on Rapid Programming Techniques. Proceedings of the Rapid Prototyping Conference, Columbia, MD, April, 1982.

12. *Requirements for Ada Programming Support Environments (Stoneman).* Washington, D.C., 1980. United States Department of Defense.

13. W. Teitelman. A Display Oriented Programmer's Assistant. CSL77-3, Xerox PARC, 1977.

14. R.T. Teitlebaum and R. Reps. The Cornell Program Synthesizer: A Syntax-Directed Programming Environment. 79-370, Cornell University, Department of Computer Science, 1979.

15. T. Wilcox and H. Larsen. The Interactive and Incremental Compilation of Ada Using DIANA. Rational, Mountain View, CA.

16. R. Yeh. "Survey of Software Practices in Industry". *IEEE Computer 17*, 6 (June 1984).

The PCTE initiative: toward a European approach to Software Engineering

Ferdinando Gallo
Bull DRTG/GL

68, Route de Versailles
78430 Louveciennes, France

This article describes the rationale for and the long term objectives of the ESPRIT project "A basis for Portable Common Tool Environments (PCTE)". The PCTE project is presented in the context of a strategic initiative which aims to establish the foundations for a systematic approach to software production in Europe. More detailed information on the technical aspects of PCTE can be found in [27, 8, 14, 15, 17, 18].

1. Summary

The project "A basis for Portable Common Tool Environments" (PCTE) has been carried out within the ESPRIT software technology area by a consortium of Bull (France), GEC and ICL (United Kingdom), Nixdorf and Siemens (Federal Republic of Germany) and Olivetti (Italy). The purpose of the PCTE project is to define and to demonstrate the usability and the implementability of a framework to serve as basis for the development of complete, modern Integrated Project Support Environments (IPSEs).

PCTE is consistent with the architectural model for Ada Programming Support Environments (APSE) described in the Stoneman report [13]. PCTE is an hosting structure which plays the role of a language independend KAPSE (Kernel APSE). The relationships between the PCTE and the ongoing Ada related initiative will be further discussed in the following sections.

The project is now completed and has achieved its principal objectives:

- PCTE is now a consolidated achievement; the technical approach has been have been validated by demonstration activities [15, 22] carried out as part of the project and by a number of external reviews and public confrontations.

- the PCTE Functional Specifications (the description of the interfaces to the PCTE services) have been published and widely distributed (in Europe as well as in the USA) [27]

- The acceptance and the credibility of the PCTE initiative is growing: a number of PCTE based projects have started already and others are being launched inside as well as outside the ESPRIT programme.

2. The need for an European Project Support Environment

The need for a systematic approach to the software production process is today widely recognised. A number of programmes have been launched in the last few years by government institutions in cooperation with industrial and academic bodies.

The trend is very much the same in Europe as in the US; it can be outlined in terms of a number of commonly accepted paradigms:

- More and more software is being required; the demand is for new services and new application areas such as networking, office automation, CAD, CAM...

- In today's commercial products, hardware technology is progressing more rapidly than software: the gap is increasing (compare for instance the evolution of microprocessors against operating systems).

Engineering techniques similar to those applying in a number of
other industrial areas (mechanical engineering, microelectronics
etc.) should be be adapted to software production to reduce the
difference between the rates of evolution of hardware and
software technologies. More powerful means of producing software
are required; these should support the activities of large teams
through the product life cycle phases.

- The effective support of formal specification techniques and of
  management of libraries of reusable software components can be a
  major factor in increasing productivity and the effectiveness of
  investments.

- Environments for the support of software project are themselves
  complex and expensive products; implementing complete project
  support environment needs very large investments both in money
  and expertise.

  One would indeed require that project support environment can
  benefit from future technological evolutions both in the hardware
  and in the software sides by preserving the original investments.

This defines the terms of the problem we are dealing with in the context of Software Engineering: the construction of very complex
software products (i.e. Project Support Environments) to boost the
productivity and quality of software production. However, no single
solution can be applicable to the construction of software for all
application domains for all organisation models and working styles.

At present, the market for Software Engineering tools is fragmented
because of a number of incompatible host environments. Writing tools
which are portable among different hosts (in order to address a larger
potential user base) implies limiting the tool implementation to use a
very narrow set of basic O.S. facilities, which can be expected to be
available on most hosts. A consequence of this approach is that the
semantics of the data manipulated by the tools is embedded in the
tools themselves; this prevents actual interoperability among tools
from different sources, though they will execute on the same machine.
Furthermore, such pratices deprive for the tool writer of the advantages of the more advanced facilities of the host environments. This
results in either lesser quality tools or in prohibitive development
costs (or both).

The awareness of these problems pushed the European information technology industry to take long term initiatives along two major axes: research in advanced technological solutions to productivity and effectiveness of software production (as epitomised by the ESPRIT Software Technology area, specifically in the family of R&D Software Engineering projects around the Portable Common Tool Environment framework) and the establishing of de facto standards as agreements among major computer manufacturers: the recent agreement between Bull, Siemens, Nixdorf, Olivetti, Philips, ICL on the X/OPEN Interfaces is an example of a (candidate) de facto standard. Five out of six of the X/OPEN group companies are also partners, with GEC (UK), in the PCTE project.

A **standards,** widely available hosting structure for project support environment would be a mean to foster investment in software engineering tools by creating a larger market for them; better tools would then become available: a multivendor, competitive market is a good way to make available to the software industry at large a comprehensive catalogue of powerful, industrial quality tools. This is the long term objective of the PCTE initiative...

Until the moment at which the critical momentum is established, the major challenge in promoting a new standard is the initial acceptance by the community to which it is addressed. The UNIX and MS/DOS phenomena can be taken, in two different domains, as two examples of what is meant here.

In the US, the KIT/KITIA (Kernel Interface Team/Kernel Interface Team for Industry and Academia) has been set by the US Departmement of Defense, as part of the STARS software initiative, to define a standard Kernel Ada Programming Support Environment (KAPSE): the Common APSE Interface Set (CAIS) [3].

On the European side, one should mention national programmes such as the Alvey (UK), the "Projet Nationale de Génie Logiciel" (PNGL, France) the "Progetto Finalizzato Informatica (Italy), the European Multi-Annual Programme (MAP) and ESPRIT.

Bodies such as Ada Europe and the ESPRIT Interest Groups have been set by the CEC to stimulate cooperation and exchange of ideas among the industrial and academic/research communities. The present European attempt to establish the PCTE interfaces as a standard basis for project support environment is, in a number of aspects, driven by the same considerations which generated the Ada and the STARS initiatives in the US; however, one can identify two major differences:

- the fundamental hypothesis is that the facilities offered by PCTE can and should be independent of the supported programming languages; language dependencies should be limited to the binding between the language and the actual facilities.

- Because of the nature of the community which was addressed, a strong requirement was set on preserving the investments already made in software engineering tools and to provide a natural, smooth path for the transition from existing pratices to the new technology. Thus, compatibility with Unix in the context of the X/OPEN group interface definition (a superset of AT&T's SVID) has been a clear mandate during the design of PCTE.

As a matter of fact, a such a standard is bound to evolve driven by advances in the state of the art; careful attention should be paid to preserve the initial investments; this condition is a prerequisite to the existence of such investments. It should however be clear that the main factor in the success of initiatives such as the CAIS and PCTE relies, above all on the quality and the adequacy of the solutions which are proposed.

The rationale for the PCTE initiative is two fold:

-- In the industrial context, the availability of good, widely accepted, public common tool interfaces is the vehicle for providing the software industry with the sophisticated tools it needs.

-- In the ESPRIT context, the notion of cooperative research imposes a clear mandate for the availability of a common environment as the vehicle for experimental implementations, dissemination and integration of results.

The ultimate goal of ESPRIT is to boost the quality and the competitiveness of the European information technology industry; in this connection, PCTE has to play its more challenging role: to be the vehicle for transfer of ideas and technology between the research and the industrial communities.

3. The foundations of the technical approach

The overal goals settled for the PCTE project (as discussed in the previous section) called for a technical approach satisfying the following principal requirements:

1) Generality: project support environments based on PCTE will be oriented towards different [software] development methods. Thus, PCTE cannot embed the knowledge about a particolar project organisation or product life cycle model; on the other hand, PCTE shuld actually support the environment designer for the definition and the implementation of specific project management/development policies.

   In particular, the generality requirement implies that the PCTE approach is independent from a given programming language, as opposed to the CAIS [3] approach, for instance.

2) Integration: it has to encourage and actually support the construction of Integrated Project Support Environment (IPSEs). The major aspects of integration are uniformity (as perceived by the environment user) and and interoperability among all the components of the environment.

3) Portability: in the context of project support environments, portability is to be addressed along three axes: portability of the IPSE itself, portability of the projects supported by the IPSE and "portability of people" (project teams and individuals).

   IPSE portability involves the ability to move the tools of the environment, from one architecture to another, as a software system.

<u>Project portability</u> involves the ability of moving, from one system to another, the project information repository (project management data, tools, documents, software configurations etc..).

<u>People portability</u> aims to minimise the cost of retraining/adaptation of personnel as they move between different systems and between different projects. The main issue to be addressed is to preserve complete <u>uniformity</u> of concepts and of the Man-Machine Interface aspect across IPSE implementations. PCTE has to ensure portability of IPSEs, project and people so that investments in IPSE technology can be cost effective and can be preserved across the evolution of hardware technology.

4) Open Endness: as IPSE technology evolves and as new areas of the software production process are investigated (for instance requirements capture and formal specification techniques), new solutions will become available in terms of innovative, more advanced IPSE components. PCTE has to support the extensibility of project support environments in a coherent, evolutionary fashion. Interoperability and uniformity has to be preserved across environment evolutions.

5) Migration strategy: the PCTE approach has to support the migration from today's software development practices to the use of IPSEs in a smooth and cost effective manner [8]. This is a vital requirement for the acceptance of the IPSE technology by the Information Technology Industry.

These requiremnts are not new; similar consideration can be found in the Stoneman [13] report and in the rationale of others studies/projects [1, 3, 11...].

PCTE can be regarded, to a large extent, as the synthesis and the consolidation of a number of earlier research and development activities. The major technical influence come from the ALPAGE [1], PAPS [11] and Unix [24,25].

The first, large scale initiatives for a systematic approach to software production (and maintenance) came from military (the Ada connection) and telecommunication organisations (the CHILL connection); both represent large and uniform communities of users with common problems related to the development and the long time maintenance of large software configurations.

The programming language is an important factor in the achievement of productivity and quality of software; however, a language alone, including a state of the art language such as Ada, addresses just the most visible aspect (and perhaps the easiest aspect, given the advanced state of the compiler construction technology) of the whole software production process. The Stoneman report advocated the need for comprehensive project support environment (the "P" of APSE is, allas, for "Programming", however, project support is clearly stated). Two concepts in the Stoneman should be considered as the foundation of the approach:

1) A Host-Target metaphor for the software production process: there is a clear distinction between the software environment support-ing the project activities (the Host) and the environment in which the software to be developed will eventually execute (the target);

2) A (traditional) layered model for the architecture of project support environments (KAPSE-MAPSE-APSEs). The role of each layer is clearly identified; in particular, a KAPSE can be regarded as a special purpose Operating System which has the responsability for supporting the integration of the various components of the [M]APSE (the tools) and for being the vehicle for the portability (of APSEs, of projects) across different architectures.

The KAPSE is the most critical component of the environment as the level of functionality and of integration which can be achieved at the APSE level very much depends on the nature of the services that are available at the KAPSE level.

Clearly, the evolution of technology from the time Stoneman was pub-lished is impressive; aspects such as personal workstations, bit map displays, Local Area Networks, Knowledge Representation systems etc. are not dealt with; however, most of the concepts are still perfectly valid.

The two concepts discussed above are also the foundation of the PCTE approach; PCTE is a Kernel Project Support Environment (KPSE, can also be interpreted as "language indepenent KAPSE) and defines a layer of services on top of which IPSEs can be built.

One should however mention the fact that some alternative architectural approaches, such as those implied in object oriented systems, can also be envisaged. However, these are still research area and a number of problems still need to be resolved in connection with the industrial usability of such techniques in the construction of IPSEs.

4. Overview of PCTE facilities

PCTE is a portable software system, however the preferred hardware architecture is a Local Area Network of powerful single-user workstations and associated resources. In this environment project teams and individual users can share software, data, and peripherals. This whole environment is seen by the user as a single machine and yet each machine can act as an autonomous unit.

Each PCTE-based IPSE is regarded as an integrated collection of tools and services specific to a particular project life cycle model and/or application domain. The public, common tool interface to the PCTE services is defined by a set of program-callable primitives which support the execution of programs in terms of a virtual, machine independent level of comprehensive facilities.

The main features of PCTE are:

--   Tool composition facilities: Execution, Communication, Inter Process Communication mechanisms.

--   The Object Management System (OMS): a fully fledged, distributed database management system based on the Entity Relationship data model, specialized for Software Engineering. The OMS defines Objects, Relationships and Attributes as being the basic items of the environment information base.

-- The User Interface: a set of facilities supporting the construction of tools based on an "object oriented" paradigm. The basic objects of the User Interface are windows, icons, (pop-up) menus, text and graphic elements.

-- Distribution: the implementation of the environment on a network of workstations includes the definition of mechanisms and protocols supporting transparent distribution of the OMS database as well as of tool executions.

-- Compatibility with UNIX: the compatibility with a widely used system in the Software Engineering domain, was identified as a major goal for PCTE acceptance and an easy migration of existing tools. PCTE extends in a compatible way the X/OPEN [24] definition of UNIX System V functionalities.

5. The future of PCTE

Today we can say that PCTE exists and is accepted in a quite large public. First of course, is the ESPRIT Programme itself: PCTE is meeting its initial challenge to be the common interface for Software Technology projects.

French and British National Projects now consider the PCTE interfaces as a basis for their future developments. The PCTE Project has also contributed to the launching of the Eureka EAST (European Advanced Software Technology) Project which is based on the PCTE interfaces.

Some European nations (France, U.K., Belgium, Italy at the time this article is prepared, others are expected to join) are endorsing PCTE as the European candidate for a NATO Interface Set standard.

A complete list of the ongoing and future PCTE related initiatives is out of the scope of this article and would probably be obsolete by the time it is published; the following is a partial picture and is mainly based on the author's involvement.

- EMERAUDE, a French National Project, involving Bull, Eurosoft and Syseca; the goal of EMERAUDE is to make available an industrial implementation of the PCTE interfaces. The EMERAUDE project played an important role in establishing the feasibility and the validity of the concepts embedded in the PCTE framework in the industrial context.

- PACT, PCTE Added Common Tools, an ESPRIT project conducted by a consortium of the PCTE and EMERAUDE partners and an Italian software house, Systems and Management; PACT aims to built a a general, method independent "minimal" IPSE. The major focus of the pact project is on Integration and open-endness. PACT will be the basis for the IPSEs built in the context of the EAST project.

- SAPPHIRE, an ESPRIT Project which aims at porting EMERAUDE to various machines and establishing a bridge between the UK Alvey Programme and ESPRIT: two Alvey Projects, ECLIPSE and FORTUNE, will migrate to the PCTE interfaces.

- EAST, an Eureka Project, aiming at constructing advanced software factories on top of PCTE. French, Danish, Finnish, Italian and British Companies are involved in EAST.

- SFINX, an ESPRIT project, which aims at validating and integrating on top of the PCTE framework the results of a number of other S.T. projects.

- ALICE, a MAP project, which aims at the design of a PCTE based distributed APSE.

A future result of the PCTE Project might be the acceptance of the PCTE Interface outside the Software Engineering world: it appears that the concepts of PCTE can be applied to CAD or office automation environments, for instance.

The Functional Specifications of the PCTE are public and are widely distributed. The current version is probably the last to be produced by the PCTE project team. The PCTE interfaces will evolve, as users will express new needs. A control body, the PCTE Managerial Board has been set; its role will be to control the evolution, to promote and support the PCTE interfaces.

6. Acknowledgements

The author would like to thank Jean Philippe Bourguignon and Ian Michael Thomas for their invaluable contributions to the form and contents of this article.

REFERENCES

[1] Atelier Logiciel Pour la Programmation des Applications de Grande Envergure, Bull, April 1982.

[2] ANSI/X3/SPARC Study Group on Data Base Management Systems: Interim Report, FDT, Bulletin of ACM-SIGMOD, Vol.7, n.2, 1975.

[3] KIT/KITIA Working Group, Proposed Military Standard Common Apse Interface Set, Department of Defense, USA,

[4] Chen, P.P., The entity relationship model: towards a unified view of data, ACM Transactions on Database Systems, 1,1, March 1976.

[5] Ichbiah J. et al., Reference Manual for the Ada Programming Language, Advanced Research Projects Agency, U.S. Department of Defense, 1980.

[6] Imperial Software Technology Ltd et all, Requirements for Software Engineering Databases, Final Report, June 1983.

[7] Kernighan B.W., Ritchie D.M., The C Programming Language. Prentice Hall. N.J. 1978

[8] Logica, ESPRIT Preparatory Study, Portable Common Tool Environment, Final Report, June 1983.

[9] Onuegbe,E., Functional Requirements for a Software Engineering Database Management System, Computer Sciences Forum, Honeywell, Vol.9 ,n.1, January 1985.

[10] Osterweil L., Software Environment Research: Directions for the Next Five Years, Computer, IEEE, April 1981, Volume 14, Number4.

[11] Portable Ada Programming System, Global Design Report, Olivetti, October 1982.

[12] Ritchie D.M. and K. Thompson, The UNIX Time-Sharing System, Comm. of the ACM 17, Vol. 7, July 1974.

[13] Requirement for Ada Programming Support Environments "STONEMAN", U.S. Department of Defense, February 1980.

[14] Knowledge Based Programmer's Assistant: Preliminary Study Report, ESPRIT Portable Common Tool Environment Project, June 1984.

[15] Knowledge Based Programmer's Assistant: Design Report, ESPRIT Portable Common Tool Environment Project, June 1985.

[16] Kennett M.R., Plan Structures of the KBPA, Document 226.1.0, ESPRIT Portable Common Tool Environment Project, 23rd. July 1985.

[17] Oddy G., The Development of the Knowledge Based Programmer's Assistant, Extended Abstract of a paper submitted to ESPRIT Technical Week 1986.

[18] Poulter K., Representing Programming Knowledge in the KBPA, Paper presented at ESPRIT Technical Week 1985.

[19] Poulter K. and Clark A., The STRATA Representation and Reasoning System, Extended Abstract of a paper submitted to ESPRIT Technical Week 1986.

[20] Rich C., A Formal Representation for Plans in the Programmer's Apprentice, IJCAI-81 vol. 2, pp1044-1052, 1981.

[21] Rich C. and Shrobe H.E., Initial Report on a Lisp Programmer's Apprentice, IEEE Trans. Software Engineering, vol. SE-4, no. 6, Nov. 1978.

[22] Functional Specifications of the CMS, Bull, GEC, ICL, Olivetti, Nixdorf, Siemens; Dec. 1985

[23] Report on the PCTE Verification Suite, Bull, GEC, ICL, Olivetti, Nixdorf, Siemens; May 1986

[24] X/OPEN Portability Guide, X/OPEN Group, July 1986

[25] System V Interface Definition, AT&T, Spring 1985

[26] X/OPEN Verification Specification, X/OPEN Group, January 1986

[27] PCTE Functional Specifications, Bull, GEC, ICL, Olivetti, Nixdorf, Siemens; June 1986

[28] C Language Test (Functional Specification), Bull, GEC, ICL, Olivetti, Nixdorf, Siemens; June 1986

[29] Pascal Compiler Validation, Brian Wichmann and Z. J. Ciechanowicz, John Wiley & Sons Ltd., 1983 Prentice-Hall, Inc., Englewood Cliffs, New Jersey

Engineering VAX Ada for a Multi-Language Programming Environment

Charles Z. Mitchell
Digital Equipment Corporation

Abstract. *DIGITAL's VAX^{TM} Ada^r is a validated, production-quality implementation of the full Ada language that is well-integrated into the VMS^{TM} operating system environment on VAX systems. The programming support environment consists of an Ada compiler, an Ada program library manager, and a multi-language programming environment including a variety of tools which all work together. The Ada compiler has many features which allow full access to the underlying operating system and hardware and which permit the use of Ada in mixed language programs in the VAX/VMS Common Language Environment. The Ada program library manager provides Ada-specific features to reduce manual bookkeeping by programmers and to support programming projects. Some of the available multi-language tools include a source code tracking system, a language-sensitive editor, a symbolic debugger, a performance and test coverage analyzer, and a test system manager.*

1 Introduction

DIGITAL's involvement with Ada parallels the history of the language itself. About the time that the *Green* language was selected and named *Ada*, we began a research and development effort that eventually resulted in DIGITAL's VAX Ada product.

During the initial stages of our research, there was some uncertainty concerning the overall thrust of the effort. The Ada language provided many new and exciting challenges in the area of compiler and run-time system design. On the other hand, the fact that Ada requires a program database seemed to open new vistas, and we would often sit back and dream about new programmer productivity breakthoughs. Was our eventual goal to produce another production quality compiler for VAX/VMS? Or was it to produce a fully integrated Ada programming environment? After several years of research and development, we realized that we weren't asking the right questions, and our actual goal emerged: to produce a production quality Ada compiler and program library manager that are well integrated into a multi-language programming environment.

---

^{TM} VAX and VMS are trademarks of Digital Equipment Corporation
^{r} Ada is a registered trademark of the U.S. Government (Ada Joint Program Office)

This paper also appeared in SIGPLAN NOTICES, Vol. 22, No. 1, Jan 1987.

2 Why Mix Languages?

The requirement to support mixed language programming results from the basic goal of producing new software at a low cost and in a timely fashion. One of the most cost-effective methods for achieving this goal is to reuse suitable existing software components, few of which are written in Ada at the current time. Thus, even when we considered the possibility of developing an Ada-only environment, we were immediately faced with the need to accommodate the use of existing non-Ada software, interfaces, and tools.

Ada will be the language of choice for many applications, and some large applications will be coded predominantly or entirely in Ada. The use of languages other than Ada may prove important even in cases where the ultimate goal is an all-Ada system. In many situations, it might be significantly more cost-effective to use some existing components that are coded in another language than to develop everything from scratch. The use of an existing numerical library, for example, might greatly reduce development risk, time, and cost. Even in cases where the long term benefits of an all-Ada system are expected to outweigh the benefits that might result from reusing existing non-Ada software, it will take many years to achieve that goal, and it is likely that many non-Ada tools will be used during the development of such applications.

Another motivation for mixed language programs stems from the belief that Ada will not immediately replace other major languages that are currently in widespread use (FORTRAN, C, Pascal, and so on). We are strong proponents of Ada and believe that the language offers significant benefits and advantages over other languages that are currently in production use. However, other languages offer advantages in some application domains, and it seems unlikely that the world's professional programming population will uniformly adopt Ada. Further, language extinction occurs at a very slow rate. People are currently relying on untold millions of lines of existing software. Although the percentage of software developed in Ada will certainly increase dramatically and will undoubtedly exceed the amount of software written in some of the currently popular languages, we expect that most existing languages will continue to be used for the foreseeable future.

Is the world ready for an all Ada (and only Ada) environment? Perhaps in the future, but not yet. We expect that non-Ada tools and components will be used for many years to produce low cost and timely software. We also expect that the ties between VAX Ada and the VAX/VMS Common Language Environment will make it easier for any installation that wants to move in the direction of Ada-only development.

3 Overview of VAX Ada

The VAX Ada product includes a validated, production-quality compiler, a program library manager, an Ada library of predefined units, and comprehensive documentation. The Ada run-time library is bundled with the VAX/VMS operating system so that a program generated using VAX Ada can run on any machine running under VAX/VMS. Symbolic debugging is provided by the VAX/VMS Debugger.

The VAX Ada compiler and program library manager can also be used in conjuction with the VAXELN Toolkit and the VAXELN Ada run-time library for embedded Ada applications designed to run on a "bare" VAX.

4 Library Manager

The VAX Ada product includes a program library manager in addition to the Ada compiler [2]. The program library manager is an Ada-only utility that reflects, to some extent, our early dreams about an improved program development environment for Ada. The language requirement to maintain a program library and dependence information facilitated the automation of a number of common programming tasks.

The program library manager, known as ACS, supports the separate compilation requirements of Ada. In addition, the program library manager helps programmers with bookkeeping and housekeeping activities and reduces the amount of manual effort required to compile and link their programs. Features are also provided to support multi-person programming projects.

The program library manager supports the standard VAX/VMS command language interface (DCL). Commands can be issued in the form of one-line DCL commands

```
$ ACS COMPILE MAIN
```

or users can enter an ACS subsystem by typing ACS at the DCL prompt ($). Once ACS responds with the ACS prompt (ACS>),

```
$ ACS
ACS>
```

any number of ACS commands can then be typed:

```
ACS> DIRECTORY A*
ACS> CHECK MAIN
ACS> LINK MAIN
```

This section briefly mentions some aspects of the program library implementation and summarizes some of the ACS features that help programmers.

4.1 Library Implementation

All units in a particular program library are recorded in a library index file which resides in a dedicated VAX/VMS subdirectory. Entries in the library index file generally correspond to compilation units and primarily contain the file specifications of various files maintained for each compilation unit by the program library manager. One such file contains an intermediate representation of the unit, as necessary to support separate compilation. Other files that are associated with each unit are the object file (machine code instructions) and optionally a copy of the source file. The location of the original source file and the time and date of compilation are also recorded in the library index file.

4.2 Concurrency and Access Control

The library index file is implemented using a multi-key, indexed sequential file, a file organization supported by the VAX/VMS Record Management Services (RMS). The file is write-shared, and RMS record locking is used to ensure that individual record reads and updates are atomic operations. When a unit has been compiled and all associated files have been written to the program library directory, the associated index file record is locked, updated, and immediately unlocked. Thus, a large number of programmers can simultaneously access a particular program library since the entire library is never locked and individual records are locked for only a very short period of time.

Access to program libraries is controlled using normal VAX/VMS protection mechanisms: hierachical protection categories (system, owner, group, and world categories with read, write, execute, and delete access) and VAX/VMS access control lists, which provide greater precision and flexibility.

4.3 Multiple Libraries

Even though the implementation permits large numbers of programmers to use and modify the same program library simultaneously, in practice it is usually desirable to insulate a multi-person project from the minute-to-minute actions of individual project members.

The program library manager allows units to be copied from one library to another. In addition, the ACS ENTER UNIT command allows units that are physically in another library to be shared by reference. For example this mechanism is used to make the predefined units in a separate system-defined program library available to user program libraries. When a user creates a new program library, the units in the predefined library are "entered" into the new library index file; that is, records containing pointers to files in the system-defined library are written to the new library index file.

Sublibraries provide the primary mechanism for controlled sharing and concurrent program development on a multi-person project. A sublibrary is statically linked to its parent at the time of creation, and a tree of hierachial sublibraries can be created to an arbitrary depth. The user can select the top-most parent or any of the sublibraries as his current program library. Units are compiled in the context of the current program (sub)library. If a unit mentioned in a WITH clause is not found in the current sublibrary, the program library manager looks first in the parent of the current sublibrary, then in its grandparent, and so on. A successful compilation modifies the current program (sub)library.

Often projects will define a parent program library that contains project software. The ACS implementation is designed so that a sublibrary can be created for each individual programmer. Each individual then works in the context of his or her sublibrary, and changes are not immediately visible to the remainder of the project. The program library manager provides a single, high level MERGE command that moves units from a sublibrary into an immediate parent and removes the units from the sublibrary. (Moving a unit involves copying the files associated with a unit and updating the library index file.) Thus, an individual can compile and test modified versions without perturbing other members of the project. Changes can be conveniently merged into the project library when appropriate.

Sublibraries can also be used for other common programming situations involving multiple versions or configurations of a system.

4.4 Building Large Programs

Efficient system building tools, similar to the UNIX[1] MAKE program, are popular because they simplify the task of compiling and linking programs. Ada requires the compiler and library manager to keep track of dependence information, and, as a result, it is relatively easy to automate the process of building programs. System building is one area in particular where it is possible to provide some Ada specific tools that are generally superior to what are usually provided for other languages.

With the VAX Ada program library manager, users can rebuild a system with minimal recompilations by typing two commands: COMPILE and LINK. The COMPILE command compares source file dates against the date of last compilation for all of the units in the transitive closure of the units specified as command arguments. The minimal set of source files are then compiled again, as necessary, so that all units in the closure are current and consistent. By default, the COMPILE command submits a batch job to do the necessary processing although other processing options are available.

The ACS LINK command checks the currency of all units in the closure of the main unit specified, writes an object file that will cause library package elaboration to occur in the correct order, and invokes the standard VAX/VMS linker.

4.5 Building Mixed Language Programs

Units compiled with VAX Ada can be linked against foreign code – that is, object code derived from non-Ada source files. This can be accomplished by specifying the names of the foreign object files as a LINK command argument. Because VAX Ada used standard object file formats, no object file conversion is required.

Foreign object code can also be introduced into a program library (either by copy or reference) as the body of an Ada library specification. The resulting foreign unit is then manipulated by the library manager as if it were an Ada unit. A typical example is a library package that declares a number of procedures, the bodies of

---

[1]UNIX is a trademark of AT&T Bell Laboratories

which are imported via pragma[2] INTERFACE. After importing the foreign code as the body of a specification, the foreign code is automatically linked into any program that depends on the associated specification.

Alternatively, Ada code can be exported and linked with non-Ada programs. The object code exported by the program library manager can be linked into a non-Ada application using the standard VAX/VMS linker. There is no requirement that the main program of a program containing Ada code be written in Ada.

5 Integrating Ada with the Common Language Environment

DIGITAL's VAX compilers are generally regarded as "industrial-strength" products with excellent performance characteristics and reliability. Another distinctive feature of DIGITAL's compilers for VAX and the VAX/VMS Common Language Environment is that a single program can be constructed from a collection of real and significant software subsystems that are written in diverse languages.

Thus, in many respects, our goal indeed was to produce yet another production compiler for VAX – this one for Ada. Given the implementation difficulties posed by Ada, our desire to produce an "industrial-strength" product was a worthy and ambitious goal by itself. In addition, a considerable engineering effort was required to make VAX Ada a full participant in the VAX/VMS Common Language Environment. This section discusses some aspects of the work that was done to provide VAX Ada users with full access to the underlying operating system facilities and to permit the use of Ada in mixed language programs.

5.1 Ada Language Features

An important integration requirement for VAX Ada was to allow full use of the VAX/VMS operating system. Chapter 13 of the Ada Standard defines implementation dependent features needed for system programming, and VAX Ada supports all implementation dependent facilities that were considered to be useful in the VAX/VMS environment [3,4]. For example, record representation specifications allow system data structures to be described in Ada. In addition, a number of capabilities allow Ada programmers to take advantage of the underlying hardware; e.g., support for three integer and four floating point types, support for VAX interlocked instructions, and so on.

5.2 Multi-Language Calls

The VAX procedure calling standard defines conventions that govern the use of basic hardware facilities, including argument list structure, register usage, stack usage, argument data types and data descriptors. Several parameter passing mechanisms and a variety of descriptors are defined. One goal of the calling standard is that a

---

[2]Compiler directive in the Ada source code

program written in any higher level language should be able to call any subprogram that conforms to the standard.

Most VAX languages pass scalar parameters and arrays with fixed bounds by reference. Thus, many mixed language calls can be made using the default parameter passing mechanisms. To allow for cases in which the called subprogram is written in a different language and expects different mechanisms, all VAX languages, including Ada, permit the caller to specify the mechanism used for each parameter.

Ada programs can call subprograms written in other languages by simply specifying pragma INTERFACE in the program source. Users can request particular parameter passing mechanisms by means of an implementation defined pragma in cases where the called non-Ada subprogram expects different mechanisms than normally chosen by the VAX Ada compiler.

5.3 Data Sharing

Although subprogram calls are usually the primary interface mechanism in a multi-language environment, users occasionally need to access variables declared in another language. This capability is supported by most VAX languages. In VAX Ada we added pragmas to permit such data sharing with appropriate restrictions to assure that language semantics are satisfied. As a result, an Ada record object, for example, can be made to correspond with a FORTRAN common block.

5.4 Condition Handling Facility

The VAX Condition Handling Facility (CHF) provides the basic facilities for error handling. Higher level language constructs such as Ada exceptions, PL/I ON units, and FORTRAN ERR= transfers are implemented using this basic capability combined with some language-specific run-time support code to handle situations in which language semantics are not identical to CHF condition handling. A result of this approach is that a condition can be signalled in code written in one language and handled in code written in another. Neither subprogram has to be aware of the programming language in which the other is written. For example the SQRT routine in the VAX/VMS Run-Time Library signals a CHF condition when requested to find the square root of a negative number. This CHF condition can be mapped to an Ada exception and handled with an Ada exception handler, even though the SQRT routine is not written in Ada.

5.5 Common Run-Time Library

General purpose VAX/VMS Run-Time Library procedures include a math library, resource allocation procedures (e.g., memory management), device-independent screen management procedures, a sort package, and so on. These utilities can be called from any VAX language.

A traditional barrier to mixed language programming has been language run-time systems that make preemptive and exclusive resource allocation decisions.

Calling a simple computational procedure written in another language only requires agreement on calling conventions and data representation. Invoking a complex subsystem written in another language requires many more agreements so that both caller and callee can, for example, perform input-output and fully utilize heap storage.

In allocating and deallocating heap storage, all callable utilities and the run-time systems for all VAX languages use routines from a hierarchy of cooperating memory allocators in the VAX/VMS Run-Time Library. Thus, a collection of multi-language routines that use heap storage can be used in the same program without fear that the run-time system of one language will interfere with that of another. Further they will use the same memory pool. For example, the same area of heap storage that is allocated and deallocated by a VAX Pascal procedure might later be allocated by a VAX Ada procedure in the same program.

It is interesting to note that prior to the release of VAX Ada, the VAX/VMS heap allocator lacked capabilities that we believed were important for our implementation of Ada. Rather than use an incompatible or uncooperating Ada-only memory allocator, we extended the VAX/VMS memory allocator to meet Ada's needs. As a result, programmers in other languages now have the benefits of sub-heaps (rather than the single heap provided previously) and a variety of allocation strategies that can be used to improve performance in many cases.

5.6 The Outcome

In integrating VAX Ada into VMS and the Common Language Environment, we encountered a number of difficulties that required creative solutions. Many of these problems stemmed from the fact that the operating system interfaces were originally defined without the benefit of a strongly typed language or that the semantics of underlying facilities differed somewhat from Ada.

One simple example of one of the many problems that had to be addressed was that many system service calls return a status value, but also update their parameters. Because such a routine returns a result, it is analogous to an Ada function; however, a function with in out or out parameters is not legal in Ada. Pragma IMPORT_VALUED_PROCEDURE was invented to allow such a routine to be interpreted as a procedure within an Ada program, and as a function in the external environment.

The VAX Ada system interfaces that resulted have proved quite satisfactory. In fact, a programmer who is experienced in Ada can quickly and easily develop a working program that makes system calls. The combination of Ada with other languages has also been quite successful. As an extreme example to indicate the degree of integration, it's possible to successfully combine a FORTRAN main program that calls an Ada subprogram that makes a task entry call to a task that calls a PL/I routine and have an Ada exception handler correctly process an ON condition raised while the PL/I routine is doing I/O.

"But wait!" the astute reader might exclaim after reading the mixed language example above. "Is it desirable to mix languages in such a manner?" The answer is yes and no, depending on the circumstances. Given a multi-person programming project, it's obviously nonsensical to allow each programmer to code in his favorite and different programming language. However, the use of multiple languages becomes more plausible and desirable when viewed in the context of reusing existing software components. A significant contribution of Ada is the ability to separate the functional characteristics of a facility from its implementation. Ada allows and encourages the development of software specifications to describe components that can be used without referring to the corresponding implementations. Thus, as long as an Ada specification fully describes the interface to a component, it need not matter to users of the package whether the implementation is coded in Ada or some other language.

As a practical example, we considered whether to use existing software components in developing the Ada compiler or to develop all software from scratch. We used various components from other products, such as the optimizer and code generator used by VAX C and VAX PL/I, after determining that this approach would allow us to produce a better product sooner and at less expense. The resulting Ada compiler is a mix of components coded in BLISS (DIGITAL's medium-level implementation language), PL/I, Macro assembler, and Ada.

5.7 A Note Concerning Program Portability

Portability was an important goal in the design of the Ada language, and Ada programs are generally more portable than programs written in other languages even if portability issues are not considered during program development. Careful attention to such issues and the use of abstraction techniques can result in highly portable programs that might require the rewrite of only a small number of package bodies to port to another system. Thus, it is possible to write Ada programs that are both highly portable and at the same time make full use of underlying features of the operating system.

As an additional aid to portability, the VAX Ada compiler provides a portability summary in the compiler-generated listing. This summary provides a cross reference to the use of all features that might affect the portability of a program. In addition, the program library manager summarizes the use of potentially non-portable features across an entire program.

6 Program Development Tools

In addition to the Ada-specific program library manager, a large number of programming environment tools are available under VAX/VMS. These provide capabilities for configuration and source management, editing, debugging, testing, and performance analysis.

A multi-language, common tool environment has many advantages for a system that supports mixed language programming. Programmers can use the same tools for different languages rather than learning a new one for each. Local programming "experts" can provide better assistance to the growing number of non-professional programmers who write programs in connection with their professional activities. For example, a multi-language debugger often allows an expert to help an engineer find a bug even though the expert may not be totally conversant in the particular programming language being used.

Some of the programming environment tools available to VAX Ada users are briefly described in this section. Operating system features are not specifically discussed. However, it should be noted that some of the simplest tools and operating system utilities are often the most useful. For example, how could one survive without a tool to search a large collection of files for a particular string? A large number of such simple utilities are available to all VAX/VMS users. In addition many aspects of the VAX/VMS operating system and command language facilities (parameterized command files, programmable command language, macro-like substitution, logical names to permit symbolic file names, file and directory search lists, useful utility commands and so on) provide the basic tools needed to handle many common configuration management issues.

6.1 User Interface

The VAX/VMS operating system and all of the VAX/VMS tools use a common interactive command language interface. Most tools are adaptable to individual tastes. Users can define their own commands, abbreviations, defaults, and screen formats. The bindings of function and keypad keys is not fixed and can be modified by users. All tools provide extensive on-line help as well as complete and comprehensive printed documentation. With the advent of the VAXstation (single-user VAX work station), the tools are increasingly providing support for work station oriented input-output, such as multi-window displays, mouse input, and so on.

6.2 Code Management System

VAX DEC/CMS (Code Management System) [5] provides capabilities similar to the UNIX[3] Source Code Control System (SCCS). Extensions beyond SCCS include additional configuration management capabilities and the ability to group collections of related files. CMS efficiently stores multiple versions of any ASCII text file by keeping only the differences between successive generations. CMS provides a file reservation system and maintains detailed historical information for individual files.

By default, the latest version of a file is retrieved when a file is fetched or reserved. An older version can also be explicitly specified. The particular versions of the files in a release or baselevel of a software system can be associated and named symbolically, such as "RELEASE_3N." It is later possible to conveniently retrieve some or all files in RELEASE_3N. Another form of grouping allows closely related files, such as all the files in the semantics analysis phase of a compiler, to be

---

[3]UNIX is a trademark of AT&T Bell Laboratories

fetched, reserved, or replaced together. CMS also provides a callable interface, thus permitting the development of site-specific source configuation control programs. Such programs have been written in VAX Ada and other languages.

CMS allows concurrent reservations and supports multiple variant lines of descent. Such support is useful in situations where work on two different versions is being done in parallel, e.g. bug fixes for the next maintenance update and new development for the next major release. CMS provides assistance in merging a variant line back into the main development path. In some cases, the merge can be done automatically, but manual intervention is required when CMS reports that multiple changes overlap each other.

CMS is a good example of a general tool that is applicable to a large number of programming languages. In addition, many people use CMS to track changes to specification and design documents.

6.3   Language-Sensitive Editor

The VAX Language-Sensitive Editor [7] is a multi-language editor designed specifically for software development. It is a multi-window, screen-oriented, programable text editor to which users can add their own definitions and templates.

One language support feature includes facilities to construct syntactically correct programs. The Editor provides templates and placeholders that simplify the process of constructing a program that is syntactically correct and appropriately formatted. Users can modify the template and placeholder definitions provided for the supported languages (Ada, BASIC, BLISS, C, COBOL, FORTRAN, Pascal, and PL/I). In addition, users can supply a definition of a language, standard form, or other textual template.

For each of the supported languages, users can construct a program by successive expansion of placeholders in templates. Placeholders are language constructs that are placed in the editing buffer as part of a template. Placeholders represent places in the source code where users must provide additional text. For example, the Ada placeholder procedure_body expands to:

```
procedure {identifier} [formal_part] is
[procedure_header_comment]
    [declarative_part]
begin
    {statement}...
[exception_part]
end [identifier];
```

Users can customize these templates to better suit the needs of their application. For example, many users redefine the template for "procedure_header_comment" to conform to local standards for code documentation.

The user can move from placeholder to placeholder or delete an optional placeholder with a single key stroke. For example, after positioning the cursor on "identifier" above, the editor will automatically replace the placeholder with the text typed by the user. Once the procedure name has been completed (by typing a space or pressing the key to advance to or delete the next placeholder), the editor automatically replaces the optional "[identifier]" at the end of the procedure with the procedure name. This latter replacement illustrates the degree of language-specific tailoring that can be achieved in a general purpose, multi-language tool.

Language specific help is provided through menus. For example, when positioned on the "statement..." placeholder, the user can easily display and select from a menu of valid choices. More detailed descriptive help is also available for all language constructs. Advanced users are not forced to use menus; they can simply type the abbreviation of the name of a template, for example "if", followed by a single keystroke to quickly get the full template for an if statement. However, advanced users can find help for seldom-used constructs quickly and efficiently.

An important question we addressed early in our editor effort was whether to develop a tree or text based editor. Although each has its advantages, the following considerations led us to adopt the text editor approach:

- There are many cases where programmers want to manipulate their programs as text, and tree editors tend to be clumsy in such situations. It's important to consider the ease with which old programs can be maintained and enhanced as well as the help provided in writing new ones.

- Users typically edit many text files other than source files and want to use the same editor for both. Programable editors are particularly desirable.

- The development of a tree to source formatter (pretty printer) that is flexible enough to be universally accepted is difficult. It's one thing for a project to mandate that a particular pretty printer be used; it's another to develop one that 100% of the programmers want to use. Many are reluctant to submit their carefully crafted code to the vagaries of a pretty printer.

- The development of a truly multi-lingual editor is difficult, especially if you want to allow users to customize it or add support for new languages.

In general, we have found that users are pleased with the language specific features provided by the editor. We find that people who have programmed in several languages can usually write a working program in a new language with no help other than that provided by the editor. New Ada programmers find that the editor helps them write syntactically correct programs. In addition, the editor-compiler interface (discussed later) gives them immediate feedback about errors and helps fix them.

6.4 Symbolic Debugger

VAX DEBUG [1] is an interactive, symbolic, multi-language object program debugger that supports a variety of languages including Ada, BASIC, Bliss, C, COBOL,

FORTRAN, Macro assembler, Pascal, PL/I, and RPG. The program to be debugged is run interactively at a terminal under control of DEBUG.

DEBUG understands multiple languages but operates according to the rules of only one language at a time. The current language is set initially to be the language of the main program, but it can be changed at any time. The debugger is fully symbolic and allows users to refer to procedures and variables using the symbolic names in the source program. Symbolic names and expressions are interpreted based on the syntax and scope rules of the current language, and output displays follow language rules.

DEBUG can also display the source text associated with the current program location. The source display window is updated each time DEBUG gets control.

In addition to standard interactive debugging capabilities (display and modification of program variables, setting tracepoints and breakpoints for subprograms, displaying subprogram call stack, and so on), a number of new features were added for VAX Ada. VAX Ada features include support for packages, WITH and USE clauses, subunits, overloaded subprograms, attributes, and tasks.

The tasking capabilities have helped users with the most frequently occurring bugs in their tasking applications. The debugger allows tasks to be referenced using program variable names or a unique identifier that is independent of the name. It provides detailed information concerning why a task is suspended and also provides a stack traceback for any task and displays variables in any task. Users can select a subset of all tasks based on priority and scheduling state and modify priorities or suspend particular tasks indefinitely. Users can also request that the debugger gain control when various unusual events occur. We expected naive Ada users to submit many tasking problem reports that would in fact be user errors. We have had very few such reports, however, since the Version 1 product was released, and we attribute the low incidence of user tasking errors to the debugging capabilities that enable users to understand the state of their tasking program. The results demonstrate how a debugger designed for multiple languages can also be very knowledgable about particular languages.

6.5 Performance and Coverage Analyzer

The VAX Performance and Coverage Analyzer (PCA) [8] is a tool used to analyze the run-time behavior of application programs. PCA can be used to identify performance bottlenecks, and it can be used for test coverage analysis.

PCA consists of two components: the collector and the analyzer. The collector gathers data from a running program and writes it to a performance data file. PCA can collect program counter sampling data, page fault data, data on calls to system services, I/O data, exact execution counts, and test coverage data. Test coverage is similar to exact execution counts, where the interesting cases are zero (not executed) and non-zero (executed at least once). Test coverage shows which code paths are never executed by a given set of test data.

The analyzer is a separate interactive program that processes the data collected during program execution and produces histograms and tabular displays of the data. The same data can be viewed in a number of ways and displayed on a terminal screen or printed. A typical use of PCA would concentrate on finding those few sections of a program that consume most of the execution time. Program counter sampling data is collected to get a general overview of program execution. Sampling incurs only a 5% overhead, so it can be used on an actual production version. PCA commands can be used to identify a hot spot, looking first at the whole program, then at routines within one compilation unit, and finally looking at lines within a single routine.

PCA is fully symbolic and uses the symbolic names chosen by the programmer. The tool can optionally display source code corresponding to a set of line numbers in its histograms and tables. For example, after using PCA to determine that a particular frequently executed subprogram is a performance bottle neck, a user can ask for a frequency histogram of program counter samples to be displayed (or printed) alongside the source. This pinpoints the particular source lines that take the most time during program execution.

As is the case with the other tools discussed, PCA can be used to analyze mixed language programs. As noted previously, the VAX Ada compiler was developed using a combination of different languages. We used PCA to analyze the performance of the VAX Ada compiler in the late stages of development. In developing the compiler we used concepts of data abstraction extensively and didn't worry about the very large number of frequently called routines we were writing. We were suprised by the execution profiles and found that substantial performance improvements were possible by focusing attention on a very small portion of the compiler.

6.6 VAX DEC/Test Manager

The VAX DEC/Test Manager [6] is a tool that helps test software during development and maintenance. This tool automates the organization, execution, and review of tests. The Test Manager is based on the concept of regression testing, in which current test results are compared against expected results. The Test Manager can be used for a wide variety of applications, including interactive applications, where the Test Manager records and replays interactive terminal sessions and verifies that the screen continues to "look" the way it should.

The Test Manager maintains a test library that consists of test descriptions and expected test results. Aids are provided to simplify the task of adding new tests. Users interactively describe their tests and assign them to user-defined groups. Multiple developers can run different combinations of tests, based on test or group names, or run the entire test suite. The Test Manager runs the selected tests, checks the results, and records the results on the test run. After the execution, the Test Manager provides a summary of results, and Test Manager users can interactively view the results at varying levels of detail.

The Test Manager is another example of a program development tool that is relatively independent of the language used to develop an application.

7 Tying the Tools Together

It wasn't all that many years ago that many program development environments consisted of a compiler and a linker. Period. Over the years additional programming tools were added, but in many cases the only relationship among the tools was that the output of one could be used as the input to another. More recently, people have come to realize that, unless a group of tools are designed to work together, their value will be less than the sum of the pieces. The importance of well integrated tools has been receiving increasing emphasis at DIGITAL and within the industry.

The previous section briefly described a number of tools on VAX/VMS but generally did not specifically address how they could be used together. This section discusses some of the ways in which these tools cooperate with each other.

7.1 Language-Sensitive Editor and Compilers

The VAX Language-Sensitive Editor allows a compiler to be invoked from an editing session with the contents of the current buffer without leaving the editing session. Processing options can be passed through to the compiler, so that, in the case of Ada, the compiler can also be invoked to do a full compilation or syntax checking only. The compiler then provides the Editor with information about any errors that were detected. Users can quickly and easily scan the errors and make corrections using the editor review mode.

In the editor review mode, the Editor splits the screen into two windows. One window displays compiler error messages and highlights the line where the error occurred, and the other window displays the source. The NEXT ERROR and PREVIOUS ERROR commands (bound to single keys) are used to move to another error in the review buffer; the GOTO SOURCE command (again, bound to a key) advances the source window to the corresponding location in the source file. Code in the source window can be modified as the errors are reviewed.

The VAX Ada compiler uses an LALR parser with sophisticated local and global error recovery mechanisms. For many source program errors, VAX Ada generates a suggested correction which it communicates to the editor. When an error is reviewed for which the compiler suggested a correction (e.g. "inserted semicolon at end of line"), the source window contains a highlighted area with the corrected text. The Editor then gives the user the choice of keeping or rejecting the correction.

7.2 Debugger and Editor

In addition to providing source display, the debugger can invoke the Language-Sensitive Editor directly when an error is discovered. The Editor opens the source file corresponding to the current program location and positions the cursor at the source line containing the error. Again, these tools can be used consistently irrespective of the VAX language being modified.

7.3   Performance and Coverage Analyzer and Editor

The Performance and Coverage Analyzer also supports source display and can invoke the Language-Sensitive Editor directly when the performance or coverage data suggest a source change. As is the case with the debugger, the editor will position the cursor at the location in the source that was being examined when the EDIT command was issued.

7.4   Ada Program Library Manager and the Code Management System

The ACS COMPILE command, in conjunction with the ACS LINK command, rebuilds a system with minimal work, as discussed previously. A search list can be used to specify directories to be searched to find the source corresponding to a unit in the program library, and a Code Management System (DEC/CMS) library can be specified in the search list. If the file is found in the CMS library, the COMPILE command checks the date-time of latest generation in the CMS library and fetches the source file for compilation if appropriate.

7.5   Test Manager and Code Management System

The DEC/Test Manager can be used with the code management system to manage files needed by the test system. Tests, test data, benchmarks, and test results can be stored in DEC/CMS libraries and the Test Manager will retrieve from CMS any files needed to run specified tests. This feature can be useful in testing multiple versions of a software system in situations where the expected results change from version to version. For example, older versions of the tests can be run and compared against results that were valid for an older, maintenance version of a system.

7.6   Test Manager and Performance Analyzer

The VAX Performance and Coverage Analyzer can be used with the DEC/Test Manager to provide information on the performance of the program being tested under various tests.

The previous section on the performance and coverage analyzer stressed the performance aspects of the tool. However, it can also be used in conjunction with the Test Manager to provide information on how well the tests cover the program being tested. It's possible to ask how well the full set of tests exercised all code paths or to check the code paths exercised by an individual test or group of tests.

8   Future Work

Integrated programming environments will eventually be the rule rather than the exception, and we expect to continue our efforts in the areas outlined in this paper. DIGITAL currently provides many of the tools needed in a full Ada Programming Support Environment, and additional tools can be expected in the future. Examples of potential future interest include support over more of the software development lifecycle and additional integration among products.

8.1 Software Development Lifecycle

Additional tools are needed to address more of the development lifecycle especially in the areas of program specification, design, and maintenance. Ada, by itself, has been successfully used as a program design language, and an Ada-based program design tool seems promising. A browsing tool that provides interactive cross referencing capabilities across a large multi-file program would particularly help in the maintenance of large programs. Many times a maintainer would like to find out where a particular variable is declared without resorting to slow, brute force techniques.

8.2 Integration Among Products

Perhaps the key to better integration is to provide better information management. In many respects, the task of providing a programming environment is one of information management. Traditionally, tools have communicated by receiving input from one tool and supplying output to others, whereas the problem might be better viewed as one of centralized data management.

9 Conclusions

The mixing of languages, including Ada, within a programming environment is important. The broad success of Ada in a variety of markets requires the gradual integration of the language into existing non-Ada environments. It must further be possible for users to leverage their investment in existing software (e.g., FORTRAN libraries) by integrating that software into Ada systems. Doing so makes not only financial sense but represents a significant productivity gain through the use of reusable software.

A common set of tools for a variety of languages, rather than language-specific tools, is highly desirable. Our experience has demonstrated that such commonality does not come at the expense of capability. It is possible to build a rich set of powerful cross-language tools. A further requirement of any tool set is that all tools be well integrated with each other.

The fact that the Ada language required a program library has promoted the development of a number of programming tools. The VAX Ada program library manager includes features for automatically building a system and other tools to automate a number of programmer activities.

The engineering of VAX Ada demonstrates that the Ada language can be closely integrated with an existing operating system and a collection of multi-language programming development tools. VAX Ada users have access to the full range of tools and facilities that run under the VAX/VMS operating system. Ada language features, including those that support data abstraction and information hiding, permit the development of highly portable Ada programs in which the use of operating system features is isolated to a relatively small amount of code.

References

[1] Beander, Bert, "VAX DEBUG: An Interactive, Symbolic, Multilingual Debugger," Proceedings of the ACM SIGSOFT/SIGPLAN Software Engineering Symposium on High-Level Debugging, *ACM SIGPLAN Notices*, Vol. 18, No. 8, August 1983, pp. 173-179.

[2] *Developing Ada Programs on VAX/VMS*. Digital Equipment Corporation, 1985.

[3] *VAX Ada Language Reference Manual*. Digital Equipment Corporation, 1985.

[4] *VAX Ada Programmer's Run-Time Reference Manual*. Digital Equipment Corporation, 1985.

[5] *VAX DEC/CMS Reference Manual*. Digital Equipment Corporation, 1984.

[6] *VAX DEC/Test Manager User Reference Manual*. Digital Equipment Corporation, 1985.

[7] *VAX Language-Sensitive Editor User's Guide*. Digital Equipment Corporation, 1985.

[8] *VAX Performance and Coverage Analyzer User's Reference Manual*. Digital Equipment Corporation, 1985.

The Ada Environment - a personal view

Vic Stenning

Imperial Software Technology Limited

Abstract. *Attention here is not confined to a programming support environment, but rather focusses on the broader notion of a project support environment. It is argued that the role of such an environment is to provide effective support for an effective process, and that this role has significant implications for the design of such environments. It is further suggested that, since few aspects of a project environment are language specific, there is little justification in producing such an environment dedicated solely to Ada. The alternative approach is to provide good support for Ada program development within a language-independent project support environment. An example is presented of one environment, ISTAR, that has adopted a language-independent approach.*

1. Introduction

Quite legitimately, the notion of an Ada† environment means many different things to many different people. Views differ, for example, on whether an environment should be dedicated solely to the Ada language or be capable of supporting a range of languages. Equally, some people advocate a programming support environment, focussing solely on the "technical" activities of program development, while others advocate a project support environment addressing broader issues such as project management and quality assurance.

Where such different views are held, it is not a case of some people being "right" and others "wrong". Rather, the differences largely reflect different concerns and priorities. The term "environment" is very widely used, but there is no accepted definition of what an environment "should do", and nor need there be. Those whose primary concern is making best use of the capabilities of the Ada language will advocate Ada-specific programming environments, while those whose primary concern is project coordination and control will probably advocate language-independent project environments. Both views are legitimate, and both are equally "correct".

Concern here is specifically with *project* support environments, rather than the narrower programming support environments. The role of a project environment is briefly discussed, and some of the implications of that role are broadly assessed. It is then argued that many aspects of a project environment are independent of the programming

---

† Ada is a registered trademark of USDOD/AJPO.

language employed by the supported project. Thus a project environment can sensibly have a basic structure that is language-independent, allowing the use of a variety of different languages. This approach offers potential advantages in terms of consistency (a single environment can be used for all tasks), mixed language working, transition into use of a new language, and economy. An example is then presented of a project environment, ISTAR, that supports a variety of programming languages, including Ada.

2. The role of a project support environment

In considering the role of a project environment it is helpful, and perhaps even necessary, to go right back to basics.

By definition, a project environment is concerned with projects as a whole, rather than with a few particular aspects of those projects. The purpose of the projects is to produce some collection of products: software, hardware, documents, whatever. (Frequently the products are not completely new, but rather are new versions or variants of some existing products.)

In any situation where projects are producing products - whether in the computer systems and software industry, or in manufacturing industry - there are certain common concerns. There is the need for the products to be of high (or at least "suitable") quality. (No attempt will be made here to elaborate on the rather complex issue of software product quality; for a discussion of the topic see, for example, section 2.4 of [1].) And there is the need for the projects to be visible, predictable, and controllable, and to produce the required products on time and on budget. Unfortunately, the software industry at least has traditionally had a rather poor reputation in some or all these areas; the need for improvement has long been recognised [2], and would still seem to be extant.

Equally unfortunately, there is no magic answer to the problems of industrial software development. Various advances - including better technical methods, "higher level" programming languages, clever tools, and a more disciplined approach to project monitoring and quality assurance - can all offer some degree of improvement, but none provides a panacea. Software development on any significant scale is inevitably complex, with many different aspects, and isolated improvements to any one aspect can yield only limited benefits.

For steady and significant improvement in the field of industrial software development, it seems necessary to focus not on one isolated aspect of the software production process, but rather on the process as a whole. Note that this process has a very broad scope and several distinct "dimensions": it encompasses, for example, all aspects of the technical life cycle from requirements capture to operational support, project management, configuration management, quality assurance, document production, and so on. Most importantly, the process also encompasses the coordination of these various "dimensions".

The objective, then, is to make advances in the practice of software development by introducing an effective process into effective use. (This, of course, is all relative; "effective" here means "more effective than before".) Note that both concerns are vital; the process itself must be effective, and it must be possible to properly employ that process in an industrial context.

Improvements in the industrial process are always slow in coming - many commentators have lamented the gulf between theory and practice - and attempts to introduce improvements often meet with resistance or failure. There are many possible reasons for this (including, for example, issues of education, scale, convenience, and familiarity), but these will not be discussed here. Instead, attention will be focussed on just one issue: the need for automated support. Experience suggests that the key to effective industrial application, whether at the level of an individual method or a complete process, can lie in good automated support. Often the barrier to industrial usage on a significant scale lies in the mass of detailed work, the sheer volume of data, and the problems of coordination. Automated support can overcome such barriers and open the way to practical application.

This then leads on to the role of an environment. It is, simply, to support the practical use of an effective process. In particular, a *project* environment must support the practical use of the *entire* process employed by the project. (Note that this role does not imply that automated support must be available for all aspects of the process. Where practical, some aspects of the process may be addressed by manual or ad-hoc procedures.)

3. Implications of the role

This role of supporting a process has a number of implications. In order to explore these it is necessary to consider certain characteristics of the process.

First, the process is both complex and varied. As noted above, it encompasses several distinct "dimensions" and many different kinds of activity: project planning, project monitoring, requirements analysis, system specification, design, implementation, operational support, theorem proving, testing, configuration management, construction and release, quality review, preparation of user guides, and so on.

Second, there is no such thing as the "best" process, any more than there is a "best" method. Rather, what constitutes an effective process is context-dependent and governed by a number of factors: the application area, development organisation, skills and experience of personnel, development priorities, and so on. Thus the process cannot be universal, but rather is tailored to a specific organisation and often to a specific project.

Third, the process will evolve. As problems with established techniques become apparent, as new techniques are developed and demonstrated to be effective, so the process will steadily change. Such evolution will certainly occur between projects, and may even occur within a given project.

These three characteristics have immediate implications for a project environment. The fact that the process is varied and complex means that the environment must support many different kinds of activity and - most important - provide coordination between these different activities. (It could be argued that the environment need not assume responsibility for coordination. However, such coordination is a critical aspect of the process, the environment should support the process, and experience suggests that manual coordination is impractical - if not totally impossible. Therefore the environment *must* promote coordination.) The fact that the process is tailored effectively means that the environment must be generic and able to support a family of processes. And the fact that the process is evolving means that the environment must have an 'open' architecture and be able to accommodate new techniques and tools.

Development of an environment that provides support for many different kinds of activity and the coordination of activities, and is also generic and open, is obviously likely to be a time-consuming and expensive undertaking. However, such an environment is inherently flexible, and most of its structure and components are independent of any particular development method or programming language. Thus the development need not be repeated for each method and language of interest; rather, it is possible to develop one common environment and then provide support for specific methods and languages within that common environment. This yields obvious advantages of economy, and also offers benefits of commonality where different methods and languages are employed, of mixed language capabilities, and in the areas of evolution and transition.

4. ISTAR - a project support environment

Previous sections have developed an argument that can be briefly summarised:

- the role of a project environment is to support an effective process

- such a process is varied and complex, tailored, and evolving

- this in turn suggests that a project environment must support many different kinds of activity and promote their coordination, and should be generic and have an open architecture

- support for particular methods and languages can then be provided within the open architecture

This section, and the subsequent sections, present an example of a project support environment, ISTAR, that follows the principles discussed above. The general structure and facilities of ISTAR are discussed, and attention is then focussed on the system's support for Ada.

The ISTAR environment has a broad scope. It supports the complete project team, rather than just individual developers. It supports all activities of all roles within the project team for the entire life cycle of the project - from initial formulation of requirements, through design and implementation, to operational usage and eventual replacement.

The design of the environment was guided by a number of basic principles. All tasks within a project should be well defined, and every member of the project team should know at all times what he or she is trying to do. There should be effective methods for addressing each identified task, and these methods should be supported by effective tools. The various tasks should be properly coordinated, so that the efforts of different team members combine to address overall project objectives. The inevitability of change should be fully recognised, and changes should be controlled, easy to make, and visible. And individual workers should be protected so that their efforts are not undermined by the actions of others, whether intentional or accidental.

The pursuit of these principles has led to an environment that supports a particular approach to systems development, which is discussed in the next section. However, it should be emphasised that this approach does not in itself represent a process in the sense discussed earlier; rather it provides a general foundation upon which support for some specific process can be built. It is by providing such process-specific support on top of the general foundations that ISTAR is tailored to the needs of particular organisations and projects.

5. The ISTAR Development Approach

5.1 The contract

ISTAR is organised to support a powerful but general approach to software and systems development - the contractual approach. This approach is based upon the hierarchical decomposition of work units into smaller work units that is typically employed for any complex project. With the contractual approach, each identified task within a project is organised as an individual contract. This contract takes as input a precise specification of the task to be performed, and produces as output the deliverables that are required from the task.

Where those responsible for a contract can identify various sub-tasks which would help to achieve the goal, and are able to precisely specify those sub-tasks, then sub-contracts can be let to perform those sub-tasks. The whole structure is of course recursive, and the sub-contracts may themselves have sub-contracts, and so on. The net result is that any task typically involves a complete hierarchy of contracts, where each of those contracts has the same basic form. Within this hierarchy, when a given contract lets a sub-contract, we refer to the former as the "client" and the latter as the "contractor".

As noted above, the main interface between a client and a contractor is that the client supplies a specification to the contractor, and the contractor returns deliverables to the client. However, for coordination purposes the client will typically need to be informed of the contractor's progress and any problems that are encountered. Further, the client may need to pass to the contractor information that is outside the scope of the specification, for example some informal response to a problem report. Thus there may be a flow of reports in both directions between client and contractor, and the complete client-contractor interface has three components: specification, deliverables, and reports.

5.2 The contract specification

A contract specification is regarded as having three distinct parts

- a task specification, which precisely defines the task to be performed (what rather than how)

- a set of verification criteria, which define an objective test to show that the task has been performed satisfactorily

- and a set of management constraints that govern the performance of the task. These may cover, for example, the required timescale for the task, resources to be employed, standards to be applied, and so on.

Of course, the nature of the contract specification will vary with the task to be performed, and different kinds of specification will be appropriate at different levels of the hierarchy. Thus, for a high level step that encompasses a major product development the task specification would typically concentrate on the user requirement and the verification criteria might call for acceptance testing according to established procedures. However, for a low level step - say one to implement the body of an Ada package - the task specification would provide the Ada package specification and a specification of the package semantics, and the verification criteria might define a specific set of tests to be performed on the implemented package.

5.3 Amendment and cancellation

It cannot be assumed that all contracts will proceed smoothly and produce their deliverables as specified and within the management constraints. Some contracts will take longer than planned, or consume more than the allocated resources. It might prove impractical or impossible to produce deliverables that meet the specification. Or the need to revise the contract specification may arise externally.

Therefore it must be possible to amend contract specifications. However, such amendments can only be made by the client. Should a problem arise within a contract then this must be reported to the client (using the normal reporting facilities) who may choose as a result to amend the contract's specification. The contractor may use reports to suggest or negotiate contract amendments, but cannot unilaterally make such amendments. By contrast, the client can make an amendment at any time - and of course must take the responsibility for doing so.

Occasionally, due to changing circumstances or insurmountable problems, it may become pointless to continue with a contract. In this case the client may choose to completely cancel the contract.

6. The ISTAR Environment

6.1 Overall structure

ISTAR provides direct support for the contractual approach, and the organisation of the system directly reflects that approach. This is perhaps most apparent in the area of the database. Rather than having a single large "environment database", ISTAR has a large number of smaller databases - one for each contract. These databases are created dynamically as new contracts are established during a project. An individual contract database acts as a repository for all information pertaining to that contract, both technical and managerial.

It is on the basis of these individual contract databases that an ISTAR environment can be distributed. Individual contract databases can be held on different machines, even when the contracts all pertain to the same project. Communication between these machines may be by means of a LAN, a WAN, point-to-point links, or even by physical transfer of magnetic media.

The major operations of the contractual approach are supported as primitive ISTAR operations. Thus there are operations for contract assignment, delivery, amendment and cancellation. The system keeps track of the machines at which contracts are located and the relationships between contracts, and any required inter-machine communication is handled automatically.

6.2 Configuration items and transfer items

Internally, an individual contract database is partitioned into a number of distinct areas. Specifically, each contract database has precisely one "contractual" area and an arbitrary number of "work" areas. As these names suggest, the contractual area is used primarily for coordination with other contracts, while the work areas are used for performing work within this contract.

Two types of information unit are particularly important within ISTAR, namely the transfer item and the configuration item. A transfer item is a single self-contained unit of information of a given type: META-IV specification, Pascal source, project plan, document, or whatever. A configuration item is a set of transfer items.

Information is held in the contractual area in the form of configuration items, and it is configuration items that are moved between contract databases. (Such moves are achieved by copying, so that following such a move the configuration item exists both in the source database and in the destination database.) Thus, when the contract is established, its specification is installed as a single configuration item in the contractual area. Similarly, any subsequent amendments are also installed as single configuration items within this area, with a relationship to the original specification and earlier amendments. And a deliverable from the contract must be established as a configuration item in this area before the actual delivery to the client can be made.

Similar arrangements apply for any sub-contracts that may be let. Thus, the specification of a sub-contract will be established as a single configuration item in the contractual area before the sub-contract is assigned. Deliverables from the sub-contract will be installed in this area as they arrive. And so on.

Once established in the contractual area, configuration items and their member transfer items are normally immutable. Work within the contract does not modify established configuration items, but rather produces new configuration items - the contract deliverable, for example, or a new sub-contract specification. This is done by first creating an empty configuration item in the contractual area, and then developing various transfer items in various work areas and "exporting" these transfer items to the contractual area. In order to produce a new transfer item it may be necessary to consult or employ some existing transfer item - from the contract specification, for example - and these can be imported into work areas as required.

Thus configuration items are held in contractual areas and moved between contract databases, while their member transfer items are imported from the contractual area into work areas, or exported from work areas into the contractual area.

Within the contractual area, both configuration items and transfer items within configuration items can exist in many distinct versions. A simple naming scheme is adopted, whereby there are distinct variant "threads" for each item, with many successors within each thread. A particular version of an item is then identified by specifying the variant and the successor, thus: STACK_SPEC(UNBOUNDED,5). Various naming defaults can then be employed when accessing existing versions, for example to access the latest version or some preferred version. Although the naming scheme is deliberately kept very simple, the data management facilities recognise a richer versioning structure, involving arbitrary trees, and record this structure by means of relationships within the contractual area. These relationships can then be queried and used where appropriate by users or tools.

The discussion thus far has perhaps suggested that configuration items can only move up or down the contractual hierarchy, between client and contractor. However this is not in fact the case. Rather, a configuration item can be moved on request from any database to any other database, subject only to access right restrictions. Such moves are normally recorded at both databases, with the source recording the destination and the destination recording the source. Thus detailed records of the movement of configuration items are automatically maintained.

This general facility for moving configuration items between databases is employed extensively within ISTAR. For example, when there is a need for a component library this is achieved by establishing a contract to operate the library. Library components are of course configuration items. New components may be submitted to the library from any source, and the source of each component is recorded. Contracts may take components out of the library as required, and all usage of a given component is again recorded. Defect reports can easily be sent to the original donor and all users of a given component, and any new version can readily be distributed to all interested users.

6.3 Workbenches

The emphasis in ISTAR is not on the individual tool, but rather on the "workbench": a coordinated set of facilities to support activities of a particular kind, such as project planning or requirements analysis. A workbench typically incorporates several distinct tools (although usually the individual tools are not directly apparent to the user) and operates in its own work area.

ISTAR workbenches fall into five distinct classes: general, technical development, project management, configuration management, and tool building.

The general class includes workbenches for text and document preparation, for timesheet completion, and for system administration. (There is also a personal mail service, but this service is pervasive throughout the system rather than associated with a particular workbench.)

The technical class includes workbenches to support the CORE method for requirements analysis, VDM for formal specification and development, and SDL (CCITT's System Description Language) for specification and design of concurrent systems. This class also includes workbenches to support various programming languages, including Ada, Chill, C and Pascal.

The project management class includes workbenches for contract management and for resource management.

The configuration management class includes a general component management workbench (covering version control, libraries, defect reporting, and so on), a build workbench, and a quality assurance workbench.

The tool building class includes a workbench that can be used to implement new workbenches to support structured development methods. Within this class there are also workbenches that address various aspects of new tool construction, including user interfacing, graphics presentation, syntax-directed editing, and database query.

7. The Ada Workbench

The majority of ISTAR's facilities are language-independent, reflecting the fact that the majority of concerns on a development project are not specific to the language that is used. Support for a particular programming language is provided by a specific workbench dedicated to that language. Thus there are separate workbenches to support, for example, Pascal and Ada. An ISTAR environment incorporating an Ada workbench provides a full APSE that meets virtually all requirements identified in Stoneman (the one significant divergence from Stoneman being that the system is not itself implemented in Ada).

As with other language workbenches, the Ada workbench exploits ISTAR's ability to accommodate existing tools without modification. Thus the Ada compiler is not written specially for ISTAR, but rather is a standard product from a commercial supplier. The first ISTAR Ada workbench is based upon the Alsys compiler, but other Ada workbenches based upon compilers from other suppliers will be developed.

The Alsys-based workbench provides facilities for syntax-directed editing of Ada source, for compilation, for basic management of an individual program library, and for program execution. This Ada workbench of course operates on an Ada work area. Within such a work area the workbench supports only a single Ada program library, enforces tight consistency constraints, and does not allow any form of versioning. This is a matter of deliberate policy. Multiple Ada libraries, versioning and configuration control are all handled in the contractual area, using the normal ISTAR facilities, rather than in the work area.

An Ada work area contains a single Ada program library and a collection of source units. The library is always consistent with the collection of source units, in that the library is the result of compiling some of those source units. However, at any time some of the source units in the work area may not actually belong to the program library. This can arise either because the unit has never been compiled into the library, or because it has been changed and not yet recompiled, or because it depends (directly or indirectly) on a unit that has been changed since this unit was last compiled.

In order to maintain the desired consistency, a unit is automatically removed from the library whenever it is edited, or whenever a unit upon which this unit depends is edited. The user must explicitly request recompilation in order to return such a removed unit to the library. Again, the explicit request for compilation is deliberate policy, primarily to provide the user with explicit control in situations where several units must be edited in order to effect a single overall change or where several editing sessions are required in order to make a lengthy change to a single unit.

Three types of transfer item are of relevance to the Ada workbench: Ada source unit, Ada library, and executable binary. An Ada source unit transfer item contains a single syntactically-valid Ada compilation unit in source form only; a transfer item of this type is exported whenever there is a requirement to preserve a source unit for future use or to transfer a source unit to some other contract. An Ada library transfer item contains a single consistent Ada library, in compiled form. The operation that exports an Ada library simultaneously exports all source units that belong to that library to the same configuration item, and the consistency of this configuration item is subsequently policed. An executable binary is a self-contained executable program.

The user interface to the Ada workbench is largely object based. Thus the user may obtain a list of all source units in the work area, or a list of those units with a given status (e.g. those that currently belong to the library, or those that have been invalidated). The user selects a unit and then selects from a menu of legitimate operations for that unit. Possible operations are edit (which encompasses simple display), compile or delete.

The Ada syntax-directed editor is a straight instantiation for the Ada syntax of the generic ISTAR syntax-directed editor. This provides the immediate advantage that editing commands are consistent with all other parts of ISTAR. The editor distinguishes between "horizontal syntax" and "vertical syntax". Broadly speaking, a horizontal syntax construct is one that would normally occupy a single line (such as a simple statement) while a vertical syntax construct is one that would normally occupy several lines (such as a compound statement). Vertical syntax constructs are automatically completed by the editor once the leading keyword has been entered, and correct vertical syntax is strictly enforced at all times. By contrast, free editing is permitted within an individual horizontal syntax construct, and no checks are made until the cursor is moved outside the construct. At this point the construct is checked for syntactic validity, and the cursor movement is inhibited if this check fails. This general arrangement offers all the normal benefits of syntax-directed editing without the irritations that arise when the approach is carried to extremes (needing to edit individual expressions in Polish form, for example).

Compilation and binding are performed by the Alsys Ada tools. The user interfacing and overall coordination of the workbench, and the import and export facilities, are provided by a top level program produced using the ISTAR tool building facilities.

Clearly, the Ada workbench is dependent upon the underlying Ada toolset. This is obviously true at the detailed level - compiler messages, detailed debugging mechanisms, and so on - but is also true at the broader level of the facilities that are offered. For example, any facility for moving units directly from one library to another (in compiled form), for merging libraries, or for building programs from multiple libraries, is dependent upon the capabilities of the underlying toolset. Our intention is to track any evolving capabilities in the Alsys toolset. When we construct other variants of the Ada workbench, based upon toolsets from other suppliers, then these variants will obviously reflect the capabilities of those particular toolsets. Nevertheless, the broad organisation and general user interface will remain largely unchanged.

8. Using the ISTAR APSE

ISTAR is intended to support the complete project life cycle, not just the program development phase. The many different activities within a project - feasibility studies, requirements analysis, prototype construction, system specification, and so on - are all conducted within ISTAR's contractual structure. These activities can be managed using the project management facilities, and their products can be controlled using the configuration management facilities.

Thus the production of an Ada program to some given specification would not typically be a complete project, but rather would be a single contract within some broader project structure. For all but a trivial program, this "Ada production" contract would itself be the root of a tree of sub-contracts, where these sub-contracts carry out different parts of the work that is needed to produce the desired program. It is natural (although not strictly necessary) for this contractual structure to largely mirror the compilation unit structure of the desired Ada program. Thus there might be one sub-contract of the overall production contract for each top level compilation unit, these sub-contracts may themselves have a sub-contract for each of their subunits, and so on. (Of course, there may be additional contracts - for example, to perform independent unit testing and QA - but these will largely supplement, rather than modify, the basic contractual structure.) Again, ISTAR's project and configuration management facilities can be used to manage these contracts and their products.

Consider a very simple program that consists of two packages and a procedure that forms the main program. One way of organising the work is to develop the two package specifications and the main program in the top level contract, but to let two separate sub-contracts for the development of the two package bodies. Thus, for example, in the top level contract the Ada workbench could be used to develop and compile the two package specifications, and the resulting library could then be exported. The library configuration item can then be used in the specification of the two sub-contracts. These two sub-contracts can subsequently deliver the respective package bodies, in source unit form, and these can be imported and compiled to complete the program.

Now suppose that one of the sub-contracts has difficulty in developing a package body to the given specification, and would therefore like a change to this specification. This sub-contract cannot make that change unilaterally; rather, it must report to the top level contract and request that the change be made. The top level contract can investigate the implications of the change, using the provided dependency tracing facilities, and decide whether it should be made. Assuming that it should, the Ada specification can be edited and the sub-contract then amended. If the second sub-contract is dependent upon the first then it too must be amended. Ensuring that the two sub-contracts are consistent is the responsibility of the top level contract; there are of course tools and facilities that assist in discharging this responsibility.

This general arrangement which mirrors the hierarchical structure of the Ada program is preferred to some "global pool of compilation units" arrangement because it is more disciplined. In particular, any interface between units is the responsibility of a single identifiable contract. It is at this contract that the impact of any potential change to that interface is assessed, and it is from this contract that the actual changes to affected units are coordinated.

It should be noted that, following the brief scenario discussed above, the history will have been recorded (including the fact that the contracts were initially let and subsequently amended). Further, the final program library and all source units will be under ISTAR configuration control. Problem reports on either the complete program or an individual unit can be submitted and tracked using general ISTAR facilities. And new versions of the program or its component units can be produced without danger of destroying or invalidating information pertaining to the existing version.

Obviously in this small scenario it has not been possible to fully illustrate use of the ISTAR APSE. The intention has simply been to sketch the general approach, and to show the way in which the general ISTAR facilities - the contractual structure, and the configuration and project management facilities - are used in conjunction with the language-specific facilities provided by the Ada workbench.

9. Conclusion

This paper has suggested that the role of a project environment is to support an effective process, and has explored certain implications of that role. It has argued that a project environment must promote coordination between a variety of different activities, and should be generic and open. Support for particular methods and languages can then be provided within the open architecture.

To illustrate the approach, an example was presented of a project environment, ISTAR, that is method- and language-independent and that emphasises coordination between the different members of a project team and the different "dimensions" of the process. The example briefly covered both ISTAR's general facilities and those that are specific to Ada.

10. Acknowledgments

ISTAR is a collaborative development by Imperial Software Technology and British Telecom. Many people at both IST and BT have contributed to its design and implementation.

11. References

[1] D. Bjorner. On the Use of Formal Methods in Software Development. Proceedings of the 9th International Conference on Software Engineering. IEEE, 1987.

Knowledge-Based Software Development
from Requirements to Code

Stephen J. Westfold
Lawrence Z. Markosian
William A. Brew

Reasoning Systems Inc.
Palo Alto, CA 94304

Abstract. This paper advocates a new paradigm for software development in which validation, modification and enhancement are performed at the level of abstract specifications. An implementation is derived from a system's specification by a process of stepwise refinement that is largely automated. We describe the REFINE™ system which supports this new paradigm including automatically compiling specifications into code. We present as an example the development of a communication system, showing in detail support for the process of producing abstract specifications from requirements.

Introduction

Software is currently the major cost in most information processing systems, and these systems are becoming more and more complex. This paper advocates a knowledge-based approach to software development environments for supporting all stages and aspects of the software life-cycle. Being *knowledge-based* means that some of the diverse logical materials of the programming process are factored out and represented formally in a knowledge base. In particular such a system should have knowledge about programs, design histories of systems, programming, and application domains in its knowledge base.

REFINE is a trademark of Reasoning Systems Inc.

[2] P. Naur and B. Randell (eds.). Software Engineering: Report on a conference sponsored by a NATO Scientific Committee. Scientific Affairs Division, NATO, 1969.

Our aim is to have the software development environment handle the details of programming, allowing the software developers to concentrate on the application problem itself and its computational nature. This level of support requires a formalization of not only the *products* of programming but also, to the extent possible, of the programming *process*. The tools of the software development environment are integrated, with the knowledge base providing the means for sharing of information. We give particular emphasis to lessening the cost of the maintenance phase of the software life-cycle because this has become the most expensive phase. A related theme is increasing the reuse of knowledge.

In this paper we first examine the software life-cycle and both the traditional paradigm for building and maintaining software and an alternative paradigm that we call the *Specification-Based Software Paradigm*. Then we present the REFINE knowledge-based system to exemplify the approach advocated and give an extended example of programming a communications system, beginning with requirements and ending with high-level program code. Finally we mention some related work and indicate some of the planned extensions to REFINE.

The Software Life-cycle

The life-cycle for large software projects is commonly divided into the following stages:

- **Requirements** gathering.
- Abstract **Specification** of the system.
- High-level **Program** development according to the specifications.
- **Validation** to ensure the specifications are met and the requirements fulfilled.
- **Maintenance** to fix errors that are discovered and to meet modified requirements.

Traditional Software Development Paradigm

In traditional software development, the requirements and specifications are used during the initial development, but maintenance and validation are performed on the program. Even if there is some effort to maintain the requirements and specification, this maintenance is separate from and incidental to the program maintenance because the relationships between the requirements, specifications and program are not available.

Studies have shown that the cost to fix an error rises exponentially with how late in the life-cycle the error is discovered [Boehm, 1981]. Given this observation, it is not surprising that as the size of software systems has been increasing, the maintenance of software is taking a larger and larger share of total project costs, so that it is now the dominant cost. We now look at one of the main reasons maintenance on high-level-language programs is so expensive.

Ideally, specifications are well-factored and problem-oriented. Program code is more oriented toward efficiency, exploiting interactions between logically separate pieces of knowledge and often using complex data structures to realize simple abstractions. For example, a solution to a problem may be concisely expressed using recursion; however, for efficiency the programmer implements it using iteration controlled by a stack. Once the decision has been made to use a stack, other uses for it may be found and it quickly becomes an integral part of the program, related to other parts in complex ways. Such optimizations spread information and introduce implementations that are efficient but complex and tangled. This exacerbates the maintenance problem by delocalizing information and increasing the dependencies among the program parts, thus making the program harder to understand.

Traditional software tools deal primarily with the *objects* of programming, such as the program text, the specifications, and the documentation, rather than the *process* of software development. Because development steps are not represented, there is no good place to document design decisions and so frequently their rationale is lost. Programmers are good at remembering such information for the short-term, but when returning to program code some time later for maintenance purposes they often have difficulty understanding even their own code and predicting the global effects of modifying it. Worse, most such knowledge accumulated by programmers is individual, and little is transferred to other project members, in particular the maintenance personnel.

The structure of the traditional paradigm for software development is shown in Fig. 1 next to the structure of the specification-based paradigm which we introduce in the next section.

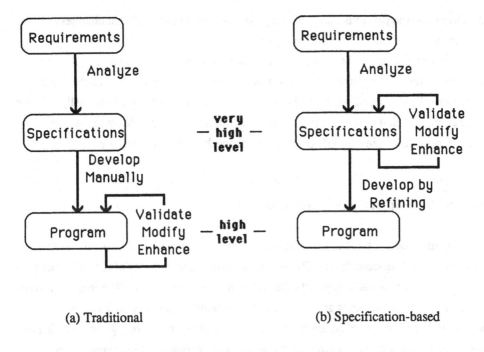

(a) Traditional   (b) Specification-based

Fig. 1. Software Development Paradigms

Specification-based Software Development Paradigm

The key feature of the specification-based software development paradigm is that maintenance is performed at a higher level, on abstract specifications, which are shorter and easier to understand than high-level-language code. This approach bypasses the problem of determining the overall effects of changes to the program, which is the biggest difficulty with the traditional approach. With this alternative approach the program needs to be redeveloped after changes to the specification. This is a more orderly problem than that of changing program code, and there is more scope for providing automated assistance. We present an automated transformational approach to solving this problem in the next section.

Knowledge-based techniques are the key to making the specification-based software paradigm practical. In order to apply these techniques to the entire software process, it is necessary to formalize the processes involved, and it is necessary to have machine capture of the process. For example, all design decisions should be recorded. This allows them to be monitored, and policy enforcement and design review can more easily be achieved. In redevelopment, the previous design history can be examined to see what changes in it are necessary with as much as possible being reused.

While our focus has been on moving activity from the program level up to the specification level, it is also desirable to continue the process by moving further up to the requirements level. Domain-specific knowledge can be brought to bear to partially validate the requirements, and to support the development of a system specification that meets the requirements. The main example of the paper shows how this can be done.

Program Development by Transformation

Program code can be developed by formal use of *automated stepwise refinement*—starting from the specifications, successive applications of formal refinement steps are made, where each step embodies a distinct design decision. The important point here is that these refinements are explicit in a knowledge base, that their rationale can be made explicit, and that there typically exists alternative refinements of a given specification. Successively choosing among these alternatives is tantamount to searching a very large space of possible implementations, a task that requires significant expertise. Since different sequences of refinements generally result in different implementations of the same specification, a relatively small number of refinements can compactly encode a wide range of implementations.

Each refinement is correctness-preserving, so the final program is guaranteed to meet the specifications without further verification (determining the applicability of individual rules typically requires simple verification tests).

The refinement of the specifications into program code can be more or less automated. At one extreme, the choice of refinements is made by the software developer, with the system recording the choices made and providing support for reusing previous refinement sequences. At the other extreme the system performs analysis to decide the most efficient choices, translating the specification into efficient target code that depends on usage and context. An intermediate position is currently taken by REFINE, with the system making the choices based on heuristics and some directives from the software developer. Because the REFINE very-high-level language can be compiled and executed, it can be used as an *executable specification language*. Software developers gain the freedom to experiment with different high-level program designs, since the task of refining a design into a procedural implementation is performed automatically by the compiler. This encourages a software development style called *rapid prototyping*, in which successive designs of a program are converted into prototype implementations, in order to provide quick feedback for validation (assessing the correctness and completeness of the design).

Several key system characteristics derive from the principle of stepwise refinement. First, a language is needed for expressing program specifications, low-level (or target-level) programs, and partially refined specifications that contain a mixed range of constructs from very-high-level to low-level. This suggests the need for a *wide-spectrum* language. Also, a language is needed for representing refinements. Refinements can be formalized in terms of *transformation rules* that rewrite one fragment of program code into another. The ability to state transformation rules succinctly requires the ability to formally express *program schemas* or *patterns*.

We next describe the REFINE system that has been built to support the specification-based software paradigm.

The REFINE Language and Environment

The REFINE language is a very-high-level programming language; it contains constructs that allow software developers to express their programs in a fashion that is easy to write, understand, and modify. These very-high-level constructs (that include operators from standard set theory and first order logic, and also a powerful state transformation operator [Westfold, 1984]) facilitate the writing of programs at the specification level. REFINE integrates features from a number of programming paradigms—object-oriented programming, logic and functional programming, rule-based systems, and conventional procedural programming.

REFINE is designed to support *self-application* [Phillips, 1983], [Green and Westfold, 1982]; since REFINE is a system for constructing programs, there is substantial leverage in applying REFINE to the task of building and extending itself. This principle has led to REFINE being *bootstrapped*—the REFINE environment is written in the REFINE language and has been used to synthesize itself. Similarly, the user can extend the system by writing new tools in REFINE and creating domain-specific representation languages.

The environment for REFINE program development includes several components, in two groups. The first group, *language-oriented components*, includes the parser, printer, compiler, type analyzer, a customized text editor, and the REFINE language itself. The second group, *development tools,* includes a command interpreter, context mechanism, documentation system, and knowledge-base browser. These components are integrated through their common use of the knowledge base. We now describe some of these components in more detail.

The *knowledge base* contains information about each construct in the REFINE language, including its syntax and the types of its arguments. It also contains the internal representation for each REFINE program that has been parsed. A facility for saving and restoring knowledge-base information from files is provided.

The *parser* converts REFINE language statements into the internal program representation used by the knowledge base. The *printer* generates the text form from the internal representation.

The *compiler* refines a very-high-level specification into a high-level language by successive applications of program transformations. Most of the transformations are specified using the pattern language constructs of the REFINE language. Patterns are used as schemas to match against the internal representation of programs in the knowledge base and to be instantiated to create the modified program representation.

REFINE is a strongly-typed language, but the burden of declarations and consistency is greatly relieved by the *type analyzer*. The type analyzer is more than a passive type checker; it combines information from user declarations, the knowledge base, and the use of data structures within the program to deduce the types of undeclared variables when enough contextual information is available to do so. The type analyzer also checks the consistency of declarations with actual usage.

The *context mechanism* maintains a hierarchy of states of the knowledge base. Previous states can be recreated. A variety of backtracking search algorithms can be easily implemented using this tool. Hence, under either program or user control, the context mechanism facilitates the consideration of a variety of program design ideas.

Example: A Communication System

The design, validation, specification and synthesis of a software system in REFINE follows several steps. These are described below for a particular example, that of building a communications system:

- **Informal requirements acquisition:** In this step, the developer states the informal functional and real-time behavioral requirements on the system, such as nature of the data to be transmitted and received, bandwidth of the channel to be used, maximum tolerable error rate, etc.

- **Requirements formalization:** The requirements are written in the REFINE language using a domain-specific vocabulary and a syntax understandable to communications engineers.

- **Requirements and design refinement:** Application-oriented refinement rules are applied to obtain successive refinements of the design, each of which contains components with increasingly detailed attributes. For example, one rule refines a transmitter into a generic noise immunization encoder if the channel to be used is noisy. Another refinement rule will refine a generic noise encoder into a particular Hamming encoder if messages are to be transmitted in packets of fixed length.

- **Design validation:** Domain-dependent and domain-independent design correctness criteria formalized in REFINE are applied to the design. For example, a check is made that no noise immunization encoding is applied prior to data encryption, and simulations are run to verify correct real-time behavior.

- **Synthesis of specifications:** Specifications are synthesized using rules that transform a domain-specific description of a component into its formal REFINE functional specification. For example, one specification expresses abstractly the output of a Huffman encoder as a function of its input, but does not provide a particular algorithm for implementing the encoder functionality.

- **Automatic program generation:** The REFINE compiler synthesizes implementations of the functional specifications automatically. The target language is an application-dependent high-level language, for example, a language for real-time programming such as Ada®.

- **Specification validation:** The target-language implementations are executed on test data. For a communications system, validation tasks include simulation of the environment, executing software on test data and analyzing the results to be sure that requirements have been met.

Ada is a trademark of the U.S. Department of Defense.

Use of REFINE in designing a communications system

In a REFINE design of a communication system, *knowledge-base object classes* are used to represent the classes of objects that occur in communications engineering applications—e.g., generic communications systems, channels, sequencing protocols, encoders, decoders, noise immunization methods, data compression techniques, encryption protocols, messages, alphabets, etc. Other object classes represent domain-independent concepts such as specification, program, project schedules, documentation, and testing suites.

Associated with each object class is a collection of *attributes* that represent properties of such objects. In the current example, standard attributes and their values for the object of class `encoder`—the object class used to represent generic data encoders—include

- `purpose`, which takes on values `noise-immunization`, `translation`, `encryption`, and `compression`,
- `input-data-type`, which takes on values `character`, `symbol`, `integer`, and `bit`, and
- `input-data-length`, which takes on integer values denoting packet size.

The syntax used to read and write the objects in an object class is defined with an extended BNF notation. User-definable syntax is important in making requirements, designs, and specifications understandable to engineers working in the application area—it allows objects to be described in the language characteristic of the application instead of a general-purpose programming language.

To provide an example of communications object classes, attributes and user-defined syntax, we focus on the first step in designing a communications system—the informal statement of requirements on its functional and real-time behavior. The requirements for our example communication system will be stated informally as follows:

- the system will transmit and receive English text messages at 4800 baud;
- the content of the messages is proprietary business information to be protected from unauthorized reception;
- the system will transmit and receive over a 9600 baud conditioned telephone line; and
- the probability of error in received English characters will be no greater than 0.00001.

Other requirements might include statements about the implementation, for example, the software is to run on a MC 68000-based processor with 1 megabyte memory, etc.

Suppose the system under development is named COMM-SYSTEM-1. Using the communications-oriented syntax defined in the Communication Systems knowledge base, these requirements can be expressed:

```
the-comm-system   COMM-SYSTEM-1
    with-message-type            English-text
    with-system-bandwidth        4800 baud
    with-probability-of-error    0.00001
    with-channel-bandwidth       9600 baud
    with-channel-characteristics conditioned-telephone-line
    with-security-level          proprietary
```

The effect of parsing this statement is to create a knowledge-base object of class "communication-system" named COMM-SYSTEM-1. The object class has attributes that represent requirements, and the particular instance, COMM-SYSTEM-1, has the indicated values for its these attributes.

In addition to representing engineering concepts in the knowledge base, REFINE can also represent *assertions* about such concepts. The following condition on correctness of a protocol is an example: a transmitter protocol is correct if and only if

1. all the encoders and decoders employed are correct;
2. the output data type of each encoder/decoder matches the input data type of its successor encoder (if the message receives further processing before going into the channel);
3. the output data type of the final encoder matches the channel , and
4. noise immunization does not precede encryption.

This correctness condition on protocol design can be represented in REFINE using the following assertion:

```
Correct(protocol) <=> Empty(incorrect-coders(protocol))
                  ∧ Empty(non-linking-coders(protocol))
                  ∧ Coder-to-channel-match(protocol)
                  ∧ Empty(noise-before-secrecy(protocol))
```

where, for example,

```
incorrect-coders(protocol) =
    {c | c ∈ encoding-sequence(protocol) ∧ ¬Correct-coder(c)}
```
and
```
Coder-to-channel-match(protocol)
    = ( output-data-type(last(encoding-sequence(protocol)))
            = data-type-of(channel-of(protocol)) )
```

These assertions illustrate the use of constructs based on logic (<=> for "if and only if", ∧ for "and", ¬ for "not") and set theory ({x | ... x ... } for forming sets, ∈ for set membership) in REFINE specifications.

In addition to their use in expressing design constraints, assertions can also test arbitrary conditions on the design knowledge base, and so provide a basis for consistency enforcement tools for project management, version control, and other software development tasks. Some constraints are so restrictive as to allow only one possibility for the value of a term once the values for the other terms of the constraint are known. For example, consider the Coder-to-channel-match constraint on protocols given above. Once the value of data-type-of(channel-of(protocol)) is known for a particular protocol then this value must also be the value of output-data-type(last(encoding-sequence(protocol))). [Westfold, 1984] describes a system that can compile constraints into efficient code for testing when the constraint is violated or for propagating values determined by the constraint.

The next step after formalizing requirements is to refine these requirements into a detailed design. The requirements themselves have already been attached to the COMM-SYSTEM-1 object. At this point COMM-SYSTEM-1 represents an abstract communications system design that is described by its requirements, but has no other associated detail. The first step in developing a detailed design is to refine the description of COMM-SYSTEM-1. REFINE has a graphics interface for displaying and manipulating knowledge-base objects that is convenient for giving an overview of the refinements. The graphics view complement the textual views which are better for concisely representing detailed information. The initial requirements are represented by a single node for COMM-SYSTEM-1, as shown in Fig. 2.

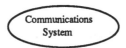

Fig. 2. Initial Requirements Node

73

The first level of refinement involves creating a transmitter, a channel and a receiver as shown in Fig. 3. This view hides the specification details of these three components that derive from the specific requirements to be fulfilled.

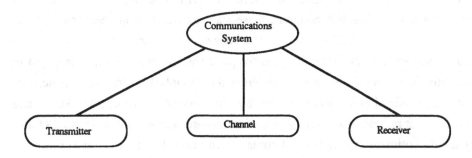

Fig. 3. System Structure after First Level of Refinement

The second level of refinement generates necessary component encoders and decoders for the transmitter and receiver respectively as shown in Fig. 4. The particular components chosen are necessary to meet the requirements. For example, a secrecy encoder is necessary because the requirements include protection from unauthorized reception. At this stage the encoders and decoders are unordered. The necessary ordering is performed by later rules which embody the knowledge that compression encoders precede secrecy encoders, secrecy encoders precede noise encoders, and the decoders are in the reverse order of the corresponding encoders.

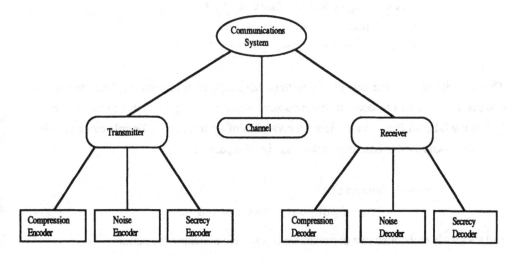

Fig. 4. System Structure after Second Level of Refinement

We now discuss one of the refinement rules in detail. The others are expressed similarly.

Knowledge about how to refine a design can be expressed in REFINE using transformation rules. A transformation is written precondition -> postcondition. REFINE compiles these state transformation specifications into executable code automatically. For example, consider the design refinement rule "if the messages are English text, then the protocol performs compression encoding." More specifically, if TRANS represents a transmitter that is embedded in a communications system that handles English text messages, then introduce a compression-encoder for TRANS that represents the production of codes—sequences of bits—from TRANS's output, with optimality rating 0.9 and compression rating 1.3, called compressor-**n** (where **n** is the first numeral not already being used to index a compression encoder). This rule is expressed in REFINE as follows:

```
TRANS = 'the-transmitter @_
            with-input-data-type @i-d-t'
  ∧ embedded-in(TRANS) = 'the-comm-system @_
                            with-message-type English-text'
->
  compression-encoder(TRANS) =
      'the-compression-encoder @newsymbol("compressor")
          with-optimality 0.9
          with-compression-factor 1.3
          with-input-data-type @i-d-t
          with-output-data-type seq(bit)'
```

This rule illustrates the use of the pattern language in transforming knowledge base objects. Patterns are enclosed in single quotes, with @ marking a variable in the pattern and "_" is a wild card (i.e., a variable whose value is of no interest). The rule can be read: if TRANS is a knowledge base object that matches the pattern

```
'the-transmitter @_
        with-input-data-type @i-d-t',
```

and TRANS is embedded in a communications system matching the pattern

```
'the-comm-system @_ with-message-type English-text',
```

then make the value of the `compression-encoder` attribute of TRANS match the pattern

```
`the-compression-code @newsymbol("compressor")
      with-optimality 0.9
      with-compression-factor 1.3
      with-input-data-type @i-d-t
      with-output-data-type seq(bit)'
```

Repeated application of domain-specific transformation rules enables successive refinement of the initial, abstract communication system into a detailed knowledge-base representation. Requirements on the overall system are refined into constraints on components that will ensure that the original requirements are met.

Transformation rules are also convenient for expressing heuristics to deal with computationally intractable problems, such as scheduling, and for system simulation using finite state machines, Petri nets, and other techniques to validate real-time behavior of embedded and distributed systems. They can also be used to automate generation of tests, documentation, and other objects associated with the specification. For example, each time a new component is added to the system design, a transformation can create an entry in a project documentation tree that includes information about the author, time of creation, details of the component, etc.

Finally, after a sufficiently detailed design is completed, transformations can be applied that generate functional specifications of the design components. For example, the following is a complete, domain-specific description of Huffman encoding, a particular technique for compression encoding:

```
the-compression-code COMPRESSOR-10
   with-unique-decodability?  true
   with-instantaneous-decodability?  true
   using-technique HUFFMAN
   with-compression-factor 0.7
   with-input-data-type seq(char)
   with-output-data-type seq(bit)
```

A transformation rule SYNTHESIZE-HUFFMAN-ENCODER-SPECIFICATION can be applied to the above description of a Huffman encoder to yield a number of specifications, including a specification of the function that encodes English text:

```
HUFFMAN-ENCODE-MESSAGE-1 (MSG: seq(char)): seq(bit) =
  [ENCODING-INT |  MSG-CHAR ∈ MSG
               ∧ ENCODING-SEQ = HUFFMAN-ENCODE-MAP-1(MSG-CHAR)
               ∧ ¬undefined(ENCODING-SEQ)
               ∧ ENCODING-INT ∈ ENCODING-SEQ ]
```

This specification uses REFINE's sequence-former construct which is analogous to the set-former construct introduced earlier in the definition of incorrect-coders. It collects all values of ENCODING-INT that satisfy the predicate into a sequence. In examining this REFINE specification we observe that the same very-high-level language that was used to validate correctness of design steps is also used to express the detailed functional behavior.

Other transformation rules applied to the description of COMPRESSOR-10 will yield REFINE specifications of related components, including a Huffman code generator and a decoder.

Finally, to give an idea of what the REFINE compiler produces, we show the derived Lisp code for HUFFMAN-ENCODE-MESSAGE-1 produced by the compiler from the above specification.

```
(DEFUN HUFFMAN-ENCODE-MESSAGE-1 (MSG)
  (LET (--SEQVAR9)
    (LET ((SF10 NIL))
      (LOOP FOR Y9 FROM 0 TO (- (LENGTH MSG) 1)
        DO (LET ((MSG-CHAR (AREF MSG Y9)))
             (LET ((ENCODING-SEQ
                     (LET ((THEPAIR9
                             (LOOP FOR PAIR9 IN
                                     HUFFMAN-ENCODE-MAP-1
                               THEREIS
                               (CHAR-EQUAL (CAR PAIR9)
                                     MSG-CHAR))))
                       (IF THEPAIR9 THEN (CDR THEPAIR9)
                         ELSE -UNDEFINED-)))))
```

```
(WHEN (NOT (UNDEFP ENCODING-SEQ))
    (LET ((SF11 NIL))
        (LOOP FOR ENCODING-INT IN ENCODING-SEQ DO
            (WHEN (NOT SF10) (SETQ SF10 T))
            (WHEN (NOT SF11) (SETQ SF11 T))
            (SETQ --SEQVAR9
                    (CONS ENCODING-INT
                            --SEQVAR9)))))))))
    (WHEN SF10 T))
    (NREVERSE --SEQVAR9)))
```

The principles underlying the REFINE compiler are described more fully in [Smith et al, 1985], [Goldberg and Kotik, 1983] and [Westfold, 1984].

Related Work

Among current research on knowledge-based programming, the transformational approach is the most widely studied. Articulation of current thought on the transformational paradigm may be found in [Scherlis and Scott, 1983] and [Möller, 1984]. A comprehensive survey of transformational systems is given in [Partsch and R. Steinbrüggen, 1983]. A complementary approach to knowledge-based programming from very-high-level specifications is based on automated deduction. In an implementation [Russell, 1983] of the logic system of Manna and Waldinger [Manna and Waldinger, 1980], the user interactively creates a proof that a given specification describes a solvable problem. A program that solves the problem is then extracted from the proof. The CYPRESS system of Smith employs "strategies" that instantiate program schemes representing various classes of algorithms, such as divide-and-conquer [Smith, 1986]. Deduction is used to derive specifications for the free operators of a scheme, thus enabling the top-down synthesis of hierarchically structured programs.

The Programmer's Apprentice project is a knowledge-based system that is not transformational. It uses plans to represent programming clichés and performs all its analysis on a dataflow graph of the system. The Programmer's Apprentice is described in [Waters, 1985]. The November 1985 issue of *IEEE Transactions on Software Engineering* is entirely devoted to knowledge-based approaches to software engineering.

A knowledge-based approach to requirements gathering using the language RML is given in [Borgida et al, 1985]. The April 1985 issue of *IEEE Computer* is entirely devoted to Requirements Engineering Environments.

In the area of languages, SETL was one of the first very-high-level languages [Schwartz, 1975]. SETL is based on the use of sets and set operations and its compiler is capable of choosing alternative implementations based on performance estimates. The CIP project in Munich developed one of the first wide-spectrum languages, called CIPL [Möller, 1984]. The CIP effort has emphasized linguistic concerns and the formalization of transformational programming. Because of its first-order logic constructs, REFINE is related to logic programming systems [Kowalski, 1979]. It differs in that generally logic programming relies on a general-purpose theorem prover (which REFINE does not). Also, in our view, logic programming systems (such as PROLOG) do not treat logic as a very-high-level language in that their compilers only use a single implementation strategy. Specification languages are widely used and several systems are described in [Cohen, 1981]. Zave and Schell describe a system called PAISLey that provides an executable specification language for modeling asynchronous parallel processes [Zave and Schell, 1985].

Conclusions

In conclusion, we have presented a specification-based paradigm for software development with knowledge-based support that promises a qualitative and quantitative improvement over the traditional paradigm. We have briefly described the REFINE system which represents a first step in building a system to support the new paradigm. We have presented a slice of software development illustrating the application of REFINE to the design, validation, and synthesis of communication systems.

We have not shown examples of maintenance, but, given a change to the requirements, a new specification can be rederived in the manner of the original derivation. Such derivation is easy because the communications knowledge that is normally only implicit in implemented communications systems has been explicated in the form of transformation rules and assertions, and is therefore readily reusable. Current work includes providing support for recording design histories and reusing them during rederivation following requirement or specification changes. We are also integrating project management support into the environment.

Specification- and knowledge-based programming environments are beginning to realize their potential for significantly increasing the productivity of software developers, allowing them to produce software that is both more reliable and more adaptable.

References

[Bauer et al, 1981] F.L. Bauer, et al, "Programming in a Wide-Spectrum Language: A Collection of Examples," *Science of Computer Programming*, vol 1, no 1, pp 73-114, 1981.

[Boehm, 1981] B.W. Boehm, *Software Engineering Economics*, Prentice-Hall, Inc., Englewood Cliffs, New Jersey, 1981.

[Borgida et al, 1985] A. Borgida, S. Greenspan and J. Mylopoulos, *"Knowledge Representation as the Basis for Requirements Specifications,"* *IEEE Computer*, pp. 82-91, April, 1985.

[Cohen, 1981] D. Cohen, "Symbolic Execution of the Gist Specification Language," *Proceedings of the Eighth International Joint Conference on Artificial Intelligence*, Karlsruhe, West Germany, pp. 17-20, 1983.

[Goldberg and Kotik, 1983] A. Goldberg, and G. Kotik, "Knowledge-Based Programming: An Overview of Data Structure Selection and Control Structure Refinement," *Proceedings of the Symposium on Software Validation*, Darmstadt, F.R.G., pp. 26-30, 1983, (Technical Report KES.U.83.7, Kestrel Institute, Palo Alto, CA, 1983).

[Green and Westfold, 1982] C.C. Green, and S.J. Westfold, "Knowledge-Based Programming Self-Applied," *Machine Intelligence* 10, Hayes, Jean E., Donald Michie and Y-H Pao, Eds., John Wiley & Sons, New York, 1982.

[Green et al, 1983] C.C. Green, D. Luckham, R. Balzer, T. Cheatham and C.Rich, "Report on a Knowledge-Based Software Assistant," RADC Report RADC-TR-195, Rome Air Development Center, Rome, New York, (also Kestrel Institute Report KES.U.83.2), August 1983.

[Kowalski, 1979] R. Kowalski, *Logic for Problem Solving*, North-Holland, New York, 1979.

[Manna and Waldinger, 1980] Z. Manna and R.J. Waldinger, "A Deductive Approach to Program Synthesis," *ACM TOPLAS*, vol. 2, no. 1, pp 90-121, 1980.

[Möller, 1984] B. Möller, "A Survey of the Project CIP: Computer-Aided, Intuition-Guided Programming—Wide Spectrum Language and Program Transformations," Technical Report TUM-I8406, Technische Universität München, 1984.

[Partsch and R. Steinbrüggen, 1983] H. Partsch and R. Steinbrüggen, "Program Transformation Systems," *Computing Surveys*, vol 15, no 3, pp 199-236, 1983.

[Phillips, 1983] J. Phillips, "Self-Described Programming Environments," Ph.D.

Dissertation, Dept. of Computer Science, Stanford University, March, 1983 (also Technical Report KES.U.83.1, Kestrel Institute, Palo Alto, California).

[Russell, 1983] S. Russell, "PSEUDS: A Programming System using Deductive Synthesis," Technical Report, Computer Science Dept., Stanford University, 1983.

[Scherlis and Scott, 1983] W.L. Scherlis and D.S. Scott, "First Steps Towards Inferential Programming," *Proc. IFIP-83*, North-Holland, New York, 1983.

[Schwartz, 1975] J. Schwartz, "On Programming: An Interim Report of the SETL Project," Technical Report, Courant Institute, New York University, New York, 1975.

[Smith, 1986] D.R. Smith, "Top-Down Synthesis of Divide-and-Conquer Algorithms," to appear in Artificial Intelligence Journal.

[Smith et al, 1985] D. R. Smith, G. B. Kotik and S. J. Westfold, "Research on Knowledge-Based Software Environments at Kestrel Institute," *IEEE Trans. Softw. Eng.*, vol. SE-11, pp. 1278-1295, Nov. 1985.

[Waters, 1985] R. Waters, "The Programmer's Apprentice, A session with KBEmacs," *IEEE Trans. Softw. Eng.*, vol. SE-11, pp. 1278-1295, Nov. 1985.

[Westfold, 1984] S.J. Westfold, "Logic Specifications for Compiling," Ph.D. Thesis, Department of Computer Science, Stanford University, June, 1984 (also Technical Report KES.U.83.8, Kestrel Institute, Palo Alto, California, 1984).

[Zave and Schell, 1985] P. Zave and W. Schell, "The PAISLey Software Tools: An Environment for Executable Specifications," *Proc. Workshop on Software Engineering Environments for Programming-in-the-Large*," Harwichport, Massachusetts, pp. 54-63, 1985.

The SMoLCS approach

to the formal semantics of programming languages

- A tutorial introduction -

Egidio Astesiano - Gianna Reggio

Dept. of Mathematics - University of Genoa - Italy

INTRODUCTION

It is now a quarter of a century since it was recognized by leading computer scientists that neither a reference manual in a natural language nor an implementation on a machine can act as a good semantic definition for a programming language. Reference manuals are usually ambiguous and can be inconsistent; implementations are too detailed and difficult to master for being used as standard references, not to say as basis of formal proofs of properties of programs.

The various techniques developed for sequential (Algol-like, Pascal-like, LISP-like) languages have shown the feasibility of precise semantic descriptions.

However, when these techniques were near to reach a satisfactory stage, it was clear that an extension was needed for coping with the problems caused by concurrency, like those arising in operating systems and in languages for handling processes in parallel.

The result has been an extensive investigation leading to significant advances in semantic techniques. But such techniques were not enough developed for handling a language like Ada during its preparation and at the moment of its official appearance in 1980. That is why many valuable attempts at a formal definition of Ada did not reach completely their aims (see [AGMRZ] for references).

The problems posed by a language like Ada are related to three main aspects:

- the strong interference of sequential and concurrent features; apparently sequential constructs, like assignment statements, may involve execution of tasks, so that it is impossible in Ada to separate sequential and concurrent aspects;
- the need of modularity due to the size of the language and to its modular structure;
- some special issues like semantics parameterized on implementation dependent features, erroneous executions and incorrect order dependences, optimization and others.

Many would argue that Ada is not the best language one could conceive on a theoretical ground having in mind good programming style and concern for correctness. However Ada is here to be used arguably for some time and an increasing number of people from the computing community are becoming concerned with the safety of Ada programs, especially because Ada was devised for embedded applications of great social relevance. Hence it seems sensible to have a good formal semantics of Ada as a starting point for the development of precise natural language explanations, of various formal documents directed to different classes of users and finally of verification methods for Ada programs; for a detailed discussion on the role of such definition (see [B]). This is the motivation for the CEC-MAP Project The Formal

Definition of Ada now under way [AGMRZ, ASP, CRAI-DDC] .

This paper presents an introduction to the SMoLCS methodology which is currently adopted in that project. However, as it will be seen, the SMoLCS methodology is not at all confined to Ada, but has as targets complex languages and large systems.

In the first section an elementary and informal presentation of the basic principles underlying SMoLCS is given, together with a short summary. Then in the second section an example of a SMoLCS specification of a concurrent system is illustrated and in the third its application to a small but significant concurrent language is shown. The paper does not give a full technical treatment which can be found in some foundational papers [AR, AR1, ARW, ARW1, ARW2].

1 BASIC PRINCIPLES AND CONCEPTS

1.1 *Denotational and operational semantics for sequential languages*

We call sequential languages the languages like PASCAL and ALGOL 60 which do not present constructs related to some kind of concurrency (shared variables, communications,...). The main techniques developed for a formal description of sequential languages are axiomatic (with Hoare-style assertions), denotational, operational and algebraic.

We are not discussing here the well-known Hoare axiomatic style; we recall instead the basic ideas of denotational and operational semantics and later of algebraic semantics. The underlying ideas of those three approaches will be helpful in order to understand our treatment of concurrent languages.

1.1.1 *Denotational semantics*

Consider a language consisting of declarations, expressions and commands, as, eg in [S]. Then the denotational semantics consists in associating a denoted value (the name) to each declaration expression and command. This is done by defining three functions called semantics functions:

D: DEC $\rightarrow$ DEC_VAL

E: EXP $\rightarrow$ EXP_VAL

C: COM $\rightarrow$ COM_VAL

where DEC, EXP and COM are the sets of syntactic objects representing declarations, expressions and commands, called syntactic domains and DEC_VAL, EXP_VAL and COM_VAL are the corresponding sets of denoted values, called semantic domains.

Now, in order to define, accordingly to our intended informal meaning, the semantic domains, we need some auxiliary structures, called auxiliary domains. For standard denotational semantics these are the domains ENV and STORE. ENV is the domain of environments, which are functions from identifiers to denoted values for identifiers:

ENV = (IDENT $\rightarrow$ DEN_VAL).

STORE is the domain of stores (or memories), which are functions from locations to storable values:

STORE = (LOC $\rightarrow$ STORE_VAL).

Then it is natural to define, for a sufficiently simple language

DEC_VAL = (ENV → ENV)

EXP_VAL = ((ENV× STORE) → STORE_VAL)

COM_VAL = ((ENV× STORE) → STORE).

Indeed, for example, a command, given an environment and a store, produces possibly a change in the store, hence another store (think, eg of an assignment).

It is convenient and has become familiar to write, for example (ENV→ (STORE → STORE)) for ((ENV× STORE) → STORE); this use corresponds to the so called "currying" technique which reduces functions of many arguments to a nested chain of functions of one argument. In the example above, the value of a command will be a function which, taken an environment, gives another function, which, taken a store, produces a store.

Thus, for example, the value of an assignment can be expressed by the clause

$$C[i := e]\rho\sigma = \sigma[E[e]\rho\sigma /\rho(i)]$$

which says that the value of the command i := e, given an environment ρ and a store σ, is a new store obtained from σ by updating the location $\rho(i)$ with the value of e, which in turn is obtained as the result of the function E applied to e, ρ and σ, ie the value of e w.r.t. the environment ρ and the store σ.

Here we adopt the usual notation: square parentheses are used, instead of round ones, around elements of the syntactic domains and we do not put parentheses around curried arguments (for example $C[i := e]\rho\sigma$ stands for $((C(i := e))(\rho))(\sigma)$).

Clearly the above clause can equivalently be written as

(∗) $C[i := e] : (\rho,\sigma) \to \sigma[E[e]\rho\sigma / \rho(i)]$

which gives explicitly the value of the command i := e.

Now a most important question arises: how do we define the functions D, E and C ? Or, equivalently, is there any standard methodology for defining them ?

The answer is: yes, by compositionality.

Informally, compositionality means that every well formed syntactic construct is given a meaning depending only on the meaning of its subconstructs and on the meaning of the syntactic operator forming that construct out of its subconstructs. We make precise the idea using again the assignment statement.

An assignment statement i := e is built using the assignment operator := and the two subconstructs i, which is an identifier, and e, which is an expression. We say that i := e is a term of type (or sort) command built using the operator := which has functionality (or type) identifier × expression→ command, where identifier and expression are the types (or sorts) of the arguments.

Then giving the semantics of i := e by compositionality amounts to say that the meaning of i := e is the result of the meaning of the operator :=, say $M(:=)$ applied to the meaning of its arguments $M(i)$ and $M(e)$; formally we should write

(∗∗) $C[i := e] = M(:=)(M(i), M(e))$.

Now $M(i)$ is just $\rho(i)$ and is a function which taken an environment ρ produces the value of the identifier (ie a location) $\rho(i)$ and $M(e)$ is just $E[e]$ and is a function which taken an environment ρ and a store σ gives as result $E[e]\rho\sigma$. Formally we can write:

(∗∗∗) $M(i) : \rho \to \rho(i)$, $M(e) = E[e] : (\rho,\sigma) \to E[e]\rho\sigma$.

Hence the above given clause $C[i := e]\rho\sigma = \sigma[E[e]\rho\sigma /\rho(i)]$ is equivalent to say that $M(:=)$ is a function which taken an element $M(i) \in (ENV \to LOC)$ and an element

$M(e) = E[e] \in ((ENV \times STORE) \to STORE\_VAL)$, gives as result a function, precisely $C[i := e]$, which, taken ρ and σ has result $\sigma[E[e]\rho\sigma/\rho(i)] = \sigma[M(e)\rho\sigma/M(i)\rho]$.

In other words, assuming $(***)$, the clauses $(*)$ and $(**)$ are two equivalent ways of defining the semantics of assignment by compositionality.

A last, most important point here is to understand that the overall set of semantic clauses, defining the functions E, D and C, is given by induction on the syntactic structure of the constructs or equivalently that we have one clause for each operator, where the meaning of the subconstructs is given in turn using the same semantic functions E, D and C (together with other known functions, of course).

As another example consider the clause for the concatenation of two commands

$C[c_1; c_2]\rho\sigma = C[c_2]\rho(C[c_1]\rho\sigma)$.

That is equivalent to the two following clauses

$M(c_1; c_2) = M(;)(M(c_1), M(c_2))$

$M(;) : f,g \to ((\rho,\sigma) \to g(\rho, f(\rho,\sigma)))$.

Summarizing, at a rather informal level, denotational semantics consists in defining a denoted value for each well formed command in a compositional way. At a bit more technical level, a denotational semantics is given by defining, inductively on the syntactic structure of constructs, a set of semantic functions, one for each type or sort of construct, after having defined the appropriate semantic domains.

Finally, in a very precise mathematical way, denotational semantics is a homomorphism from the algebra of syntax into a semantic algebra (it is well known that this homomorphism is unique); the algebra of syntax consists just of the syntactic constructs with their structures, the semantic domains are the semantic carriers of the semantic algebra, the meaning of the operators are the interpretations of the operators in the semantic algebra and the semantic functions are the functions constituting the homomorphism.

1.1.2 *Operational semantics*

Operational semantics consists in giving a set of rewriting rules (a calculus), transforming a program into its result; non-termination is usually indicated by an endless sequence of rewritings.

For many years every operational semantics was associated to a more or less abstract machine for executing programs. For example, the cycle consisting of the execution of a program on a real machine (compiler, assembler,...) can be seen as the definition, at a very low implementative level, of an operational semantics. However this kind of semantics is useless for the purposes we have seen in the introduction (standard reference, correctness checks,...). That is why soon it became clear that higher level operational description were needed; one of early example was the PL/1 semantics developed in Vienna, based on an underlying abstract machine.

However a real breakthrough was provided by the Structural Operational Semantics (abbreviated to SOS) advocated by Gordon Plotkin. SOS has been used by Plotkin and others starting in the mid-seventies, but its explicit formulation has to be found in Plotkin's lecture notes [P] and in its notable applications to CSP by Plotkin and to CCS by Milner [M1].

The principles of SOS are simple and can be better understood now, having in mind the principles of denotational semantics. Indeed the two techniques share the idea that the semantic definition is given by structural induction on the syntactic structure of terms (hence the name SOS) and as a consequence the basic tool for proving properties derived by the definition is proof by structural induction. Then the difference of SOS with respect to denotational semantics is the use of rewriting rules instead of assignment of denoted values.

For example the assignment statement is handled by the clause

$$\frac{<e,\rho,\sigma> \xrightarrow{\quad E \quad} v}{<i := e,\rho,\sigma> \xrightarrow{\quad C \quad} \sigma[v/\rho(i)]}$$

where $\xrightarrow{\quad E \quad}$ and $\xrightarrow{\quad C \quad}$ are two binary (rewriting) relations, the first for evaluating expressions and the second for evaluating commands. The clause is an inference rule, saying that if the expression e evaluates to v in the environment ρ and with the storage σ, then the command i:=e, under the same ρ and σ, evaluates to $\sigma[v/\rho(i)]$.

The analogy with the corresponding denotational clause is clear as much as the inductive structure of this definition, which uses $\rho(i)$ and $\xrightarrow{\quad E \quad}$ for evaluating the subconstructs i and e.

The only tricky point in comparing the two techniques is how they handle nontermination and can be explained by discussing the concatenation of two commands, say $c_1 ; c_2$.

If the evaluation of c_1 does not terminate, its value $C[c_1]\rho\sigma$ in denotational semantics is a special value $\perp$ and then since $C[c_2]\rho(\perp) = \perp$, the whole command has $\perp$ as value (see the clause above).

In the SOS style we have two main possible choices, expressed by two sets of clauses:

$$\text{PD} \quad \frac{<c_1,\rho,\sigma> \xrightarrow{\quad C \quad} \sigma', \quad <c_2,\rho,\sigma'> \xrightarrow{\quad C \quad} \sigma''}{< c_1 ; c_2,\rho,\sigma> \xrightarrow{\quad C \quad} \sigma''}$$

$$\text{NT} \quad \frac{<c_1,\rho,\sigma> \xrightarrow{\quad C \quad} <c_1',\rho,\sigma'>}{< c_1;c_2,\rho,\sigma> \xrightarrow{\quad C \quad} < c_1';c_2,\rho, \sigma'>} \qquad \frac{<c_1,\rho,\sigma> \xrightarrow{\quad C \quad} \sigma'}{< c_1; c_2,\rho,\sigma> \xrightarrow{\quad C \quad} <c_2,\rho, \sigma'>}$$

The clause PD implies that we cannot derive any rewriting sequence for $<c_1; c_2,\rho,\sigma>$ in the case $<c_1,\rho,\sigma>$ does not evaluate to any state σ'; the set of clauses NT implies that a nonterminating sequence of rewritings for $<c_1,\rho,\sigma>$ results also in a nonterminating sequence for $<c_1; c_2,\rho,\sigma>$.

Hence the first kind of clauses represents nontermination via partial definedness (hence PD): a construct whose evaluation does not terminate has no value; the second kind of clauses represents nontermination via a nonterminating sequence of intermediate steps (hence NT). The second approach is the only appropriate when also nonterminating evaluations produce values, like an endless sequence of printed messages. We will see that this second approach is also appropriate for handling concurrent programs.

1.2 *From sequential to concurrent languages*

We have seen that in the sequential case the semantics of a construct is a function in the denotational approach and a rewriting sequence in the operational. Note now that a rewriting sequence can be viewed as a transition sequence in which the rewriting of a term t into a term t' can be seen as a transition from a state t to a state t'. It turns out that this interpretation of the operational semantics approach can be extended easily and nicely to give a semantics of concurrent languages, later in this section we shall see that also the functional approach can be seen as a specialization of this extension.

1.2.1 *Labelled transition systems as models of processes*

A most fruitful idea, which has found an extremely elegant application in CCS and SCCS [M1, M2], widely considered a major breakthrough in the theory of semantics of concurrency, is that a process is specified as a labelled transition system and modelled as a labelled, possibly infinite, tree.

A labelled transition system is a set of triples (s,f,s'); a triple is also written $s \xrightarrow{f} s'$ and means that the system passes from the state s to the state s' under an interaction with the external environment represented by the label (or flag) f. In the simplest case when the transition is purely internal to the system and there is no relationship with the environment, the label can be dropped or better represented by a special label, which is usually TAU (or τ) as it is in CCS.

It is quite important to understand that there are two interesting views of f: one is of f as a condition on the environment for the transition to be performed; the other is of f as what can be observed from outside of the system when it performs a transition. Both these points are well discussed by Milner in [M1] and also illustrated in more detail by later examples.

Given a transition system, to each state s is associated a labelled tree, in which we do not care of permutation of branches and two equal subtrees with the same root are considered once (in the following when speaking of labelled trees we will always consider them up to this equivalence)

Branching on labelled trees represents nondeterminism. Let us show why nondeterminism arises in a process. A typical example is a process modelling a memory register which can either be written or read; this can be formalized by a process in a state s which can either receive a value x or send out a value v ; a possible representation of s as a term is

$$s = \textbf{choose } IN(x) \; \Delta \; s_1 \textbf{ or } \; OUT(v) \; \Delta \; s_2$$

where **choose ... or...** represents a nondeterministic choice, Δ means "followed by" and IN(x), OUT(v) corresponds to the two actions of receiving and sending.

The corresponding representation of s as a tree is

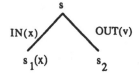

The above description of a memory register is not complete; indeed while v is just the value stored in the register, the value x which can be written can be any value in the range of values, say the set of naturals

NAT. Hence a more appropriate representation would be

$s =$ **choose** $x :$ NAT $\quad$ IN(x) Δ s_1 $\quad$ **or** $\quad$ OUT(v) $\quad \Delta s_2$

where **choose** $x :$ NAT means clearly a choice for x in NAT; graphically this can be represented as

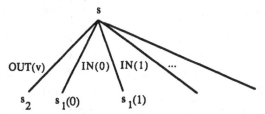

In other words an action parameterized on a value like IN(x) for x in NAT, is represented by an infinite branching tree, with one branch for each value.

Moreover, a complete formalization of a memory register holding a value v, would be, see [M1],

$\qquad$ MREG(v) =def= **choose** $x :$ NAT $\quad$ IN(x) Δ MREG(x) **or** OUT(v) Δ MREG(v)

which shows a recursive definition. Later recursion will be defined by a fixpoint operator **fix**.

We have seen how nondeterminism arises from an intrinsic multicapability of actions. But nondeterminism can also derive from parallelism, when a process can perform an action with at least two other processes, for example when a process has the ability of performing a call and there are in parallel two other processes able to accept that call; clearly there are at least two mutually exclusive possible transitions of the system in that situation.

In conclusion we have seen how branching on labelled trees modelling processes comes out. Later in section 2 this idea will be made technically precise and also it will be shown how the cooperation among processes can be formally expressed by operations on labelled trees.

1.2.2 *Semantics of processes*

Representing a process as a labelled tree is already a move towards giving a semantics to a process; indeed this corresponds, as we have already pointed out, to extend the operational semantics to handle concurrency.

However the difficult problem in semantics is abstraction, ie to give a semantics sufficiently abstract w.r.t linguistic details. In this respect while labelled trees are much more abstract than pieces of code, it is nevertheless clear that in many cases there are different trees with the same semantics. One of the simplest cases is when we consider processes corresponding to purely sequential commands; their models as labelled trees are unary trees with all the arcs labelled by the symbol of internal action TAU, like in figure 1.

Then, if we are interested in an input-output semantics, we would say that the two trees (here sequences) are equivalent iff s_I and s_I' are input-equivalent and s_F and s_F' are output-equivalent, irrespective of differences in other aspects like the intermediate states.

From this simple example we can understand that a semantics is then given by an appropriate equivalence on labelled trees. The case we have just discussed informally shows that functional semantics for

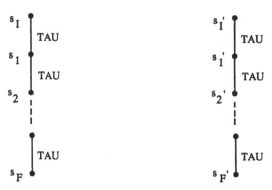

Figure 1

sequential languages can be obtained as a special case of semantics for concurrent languages, just considering two trees (now just unary trees as those seen above) semantically equal whenever they are input-output equivalent.

Thus it is easily understood that other interesting equivalences on trees can arise, depending in general on what we want to observe of a system. For example two well known equivalences are strong-equivalence and stream equivalence. For the first we consider two trees equivalent iff they are the same tree by forgetting the states (always modulo the equivalence defined in the beginning). For the second two trees are equivalent iff the two corresponding sets of label sequences starting from the two roots are the same. Later on we will see that in SMoLCS it is possible to assign a semantics depending on what we want to observe, following a precise parameterized schema.

1.3 *Abstraction and modularity*

1.3.1 *Formal semantics versus specification and verification*
A basic requirement of a formal semantic definition is abstraction or, equivalently, independence from implementation details. However this requirement has to be considered together with a set of generally accepted principles which in some sense pose some precise constraints on the level of abstraction of a formal definition. The first unquestionable principle is for a FD to keep the language syntax; otherwise a user would not recognize the language. Of course the syntax can be given a somewhat abstract form, but in any case the link with the concrete syntax should be clearly established. A second widely accepted principle is compositionality, in the sense we have explained in the preceding sections. Altogether these two principles seem in some contrast with an abstract semantics based on the specification of some higher level properties of a language. But the simple fact is that a formal semantics for a programming language cannot be confused with the formal specification of a software system; in this second case there is no syntax around, our specification establishes both syntax and semantics and clearly there is an advantage to keep the level of specification as higher as we can. For example our specification can be compositional, but the level of the components can be chosen at the highest granularity level permitted.

For a formal semantics the pattern for achieving abstraction can be the following: start from the language syntax, give a compositional semantics in which every auxiliary structure is abstractly specified, define an inducted semantic equivalence on syntactic well formed constructs and then derive from the formal semantics higher level properties (now theorems about the FD) to be used in verification and correctness proofs. This was more or less the pattern already followed in classical denotational semantics, see [S].

1.3.2 *Abstract data types*

A useful tool for achieving abstraction is the use of abstract data type (abbreviated to adt from here on) techniques in formal semantics.

By an abstract data type specification we mean a signature and a set of axioms. A signature is a set of operation symbols with their functionalities; usually we write

$$f: s_1 \times s_2 \times ... \times s_n \rightarrow s$$

for saying that s_1, ..., s_n are the sorts of the arguments and s is the sort of the result.

Since we use a partial data type approach [BW2], ie the value of an operation can be undefined for some arguments, also definedness predicate symbols are used, one for each sort, to say that an object is defined; all are indicated by D (the sorts can be deduced from the context).

Axioms are always first order formulas in positive conditional form, ie of the form e **IF** $\wedge$ e_i, where

e can have form either $D(t)$ or $t_1 = t_2$ and e_i either $D(t_i)$ or $(D(t_i) \wedge t_i = t_i')$. It is assumed that every axiom is implicitly universally quantified over all variables, but variables can only range over defined values in the interpretation. Terms and axioms are interpreted in partial algebras, which are structures consisting of a set of carriers, one for each sort in the signature, and of a set of functions corresponding to the interpretations of the operation symbols (including the definedness predicates, which are assumed to be total, ie either true or false on every element).

There are two other important points about interpretation: first $t_1 = t_2$ is true iff either both t_1 and t_2 are defined and equal or both are undefined; second the functions are strict, ie if $f(t_1,...,t_n)$ is defined, then all $t_1,...,t_n$ are defined.

Let us now explain the use of abstract data types techniques in formal semantics.

Obviously such techniques can be applied to the specification of the data structures defined in the language and of those used for defining auxiliary structures, like memories environment, denotations and so on.

It is less clear how those techniques can be applied to give an overall formal definition of a language as the specification of a data type. This specification was pioneered by the Munich CIP group, especially by Broy and Wirsing, see [BW3, CIP, BW1].

Later we will see how the SMoLCS methodology makes an extensive use of adt techniques; in particular the specification of a labelled transition system will be just the specification of an abstract data type (see [ARW1]).

Together with abstraction, modularity is the other major benefit of using such techniques and clearly modularity is a basic requirement for the specification of complex systems and languages. In our approach modularity has two facets: one is putting together module specifications, the other is parameterization. All the SMoLCS approach is highly modular and parameterized; note that, parameters can be entire specifications with their sets of axioms. It is for example this ability which allows to specify a semantics for Ada, which is intrinsically parameterized on implementation dependent features.

1.4 An informal overview of SMoLCS

In this section we give a very informal overview of the SMoLCS approach, which now should be understandable because of the concepts given in the previous section. Then in sections 2 and 3 a more technical, though still introductory, presentation of SMoLCS will be given.

1.4.1. General structure of SMoLCS

SMoLCS is an integrated methodology for the specification of concurrent systems and languages developed mainly in Genova (Dept. of Math.) by the authors (see [AR, AR2]), and relying for the algebraic setting on a cooperation with M.Wirsing (Passau - Fakultät für Informatik) (see [AMRW, ARW]).

The typical field of application of SMoLCS are large systems, multilevel architectures built from systems with different granularity, complex concurrent languages with modules and interference between sequential and concurrent features.

For the specification of concurrent systems, SMoLCS has been applied to specify the internode communication architecture of the project of Cnet (a local set of workstations) (see [AMRZ], [AMRZ1]).

As a method for the specification of languages, it is the methodology chosen for the formal definition of the dynamic semantics of Ada in the CEC-MAP project, see [AGMRZ], [CRAI-DDC]).

SMoLCS consists of four parts: **specification of concurrent systems, semantics of concurrent languages, metalanguage, tools** (under development).

Its main features are the following.

SMoLCS combines algebraic, operational and denotational techniques.

The **specification of a system** is obtained as an instantiation of a parameterized data type, following a schema based on an operational intuition of a process as a labelled transition system and of a concurrent system as resulting from the composition of the subsystems.

The abstraction from the operational intuition is obtained by a schema ensuring the existence of an observational semantics, represented by an algebra.

The **semantic definition of a language** is compositional and adopts a denotational style, but can support the algebraic specification of modules. Moreover both classical styles used in denotational semantics, the Oxford continuation style, and the VDM direct semantics style are supported (and even shown to be a variant of a unique formalism see [AR1]).

The **metalanguage** consists of two kernels, one applicative and one algebraic, with a simple semantics consisting in a simple connection between the semantics of the two kernels, which are given following classical methods. Moreover the metalanguage can be mapped completely into the high-level algebraic specification language ASL ([SW, W, AW]).

Many well known formalisms for concurrency, like CCS, SCCS and CSP, TCSP can be easily defined in SMoLCS. More importantly we can obtain as a special instantiation of SMoLCS an extension of VDM which admits algebraic specifications and can handle concurrency, still using some of the basic combinators of the direct semantics style, typical of VDM (see [AR1]).

The **tools** under development are mainly aimed at the production of a symbolic parameterized (nondeterministic) interpreter of SMoLCS, which, combined with a processor of denotational clauses, will make executable the formal definition of a language (Ada, for example). It corresponds, more or less, to an interpretation for the metalanguage, apart from the observational semantics. Until now a specific rapid prototyping tool has been realized for SMoLCS ([Mo]) which is a variation of the RAP system [H], specially tailored to the structure of SMoLCS. It consists of a concurrent symbolic interpreter, which can derive transitions for a specified concurrent system, and of an interpreter for denotational clauses.

Let us now discuss the first two parts in more detail.

1.4.2 *Specification of concurrent systems*

A concurrent system is a labelled transition system built from some component subsystems; each subsystem is in turn modelled as a labelled transition system.

A state of a concurrent system is modelled as a set of states corresponding to the subsystems plus some global information, the transitions are inferred from the transitions of the component subsystems in three steps: synchronization, parallelism, monitoring.

- Synchronization defines the transitions representing synchronized actions of sets of subsystems and their effects on global information.

- Parallelism defines the transitions representing admissible parallel executions of sets of synchronized actions and the compound transformations of the global information (mutual exclusion problems, eg, are handled here).

- Monitoring defines the transitions of the overall system respecting some abstract global constraints (like interleaving, free parallelism, priorities, etc...).

This SMoLCS schema is expressed in an algebraic parameterized way so that every instantiation on the appropriate parameters, defining the information for synchronization, parallelism and monitoring, is an abstract data type .

More precisely the definition of a SMolCS specification of a system is modular, hierarchical and parameterized.

Every composition step is a parameterized abstract data type specification: for example the synchronization step STS takes as parameters the specifications of a basic transition system BTS, a global information structure specification INF, new flags defined by a specification SFLAG and axioms E_{SYNC} (which are instantiations of a unique axiom schema, typical of synchronization) and give a labelled transition system $STS(BTS, INF, SFLAG, E_{SYNC})$ which is an instantiation of STS on the above parameters.

Analogously for the other steps PTS (for parallelism) and MTS (for monitoring).

As a result the overall SMoLCS schema can be seen itself as a parameterized abstract data type SMoLCS, such that, given a basic transition system BTS and the appropriate parameters for each step, the specified system SYST is just

$$SYST = SMoLCS(BTS,...) = MTS(PTS(STS(BTS,INF,SFLAG,E_{SYNC}),...)),$$

ie the hierarchical composition of the three steps.

Clearly, since a system like SYST is the specification of a labelled transition system, it can be taken as a basic transition system for another SMoLCS specification; that means that the SMoLCS construction can be iterated. Moreover it is possible to give a SMoLCS schema corresponding to an inductive definition.

A pragmatic advantage of this parameterized approach is that it is enough to give the user the directions for the specification of the parameters. An appropriate friendly syntax has been developed for encouraging structured and correct specifications. Moreover some rules are given ensuring the existence of initial models for each specification.

Together with an initial algebra semantics, corresponding to an operational semantics, the SMoLCS approach supports, with explicit linguistic constructs, the definition of an observational semantics again via a parameterized abstract data type specification, where the parameters corresponds to a formalization of the observations. Every instantiation of such schema admits a terminal model, the Concurrent Algebra, in which two states of the concurrent system are equivalent if and only if they satisfy the same observations; moreover every subcomponent of the state gets an observational semantics by closure with respect to state contexts (see [ARW] for foundations). Note that this is just an existential definition, to guarantee consistency; for any instantiation such observational semantics has to be characterized more explicitly, by suitable equivalences on the labelled trees associated to the states and subcomponents. The above schema permits to formalize observationally semantics like input/output, strong equivalence, classes of bisimulation equivalences, streams and test semantics.

1.4.3 *Semantics of concurrent languages*

The SMoLCS methodology is based on a two-steps approach, combining a denotational overall schema with algebraic techniques for the specification of abstract data types, here applied to the specification of concurrent systems, which are formalized following the SMoLCS operational schema (see [A, AR, AMRW, ARW]).

Essentially the first step connects the abstract syntax to an underlying model for concurrency, formalized in a suitable language for describing processes (behaviours) and their mutual interactions in a concurrent system. This is done by a set of <u>denotational clauses</u>, where in a typically denotational style to each well formed construct of the source language a term in another language is associated. For example a function like

$$P : PROGRAM \rightarrow STATE$$

will associate to a program a term representing a state of concurrent system corresponding to the execution of the program.

A function like

$$E : STATEMENT \rightarrow (LOCAL\_INF \rightarrow BEHAVIOUR)$$

will associate to a statement, given some local information (having a role similar to the environment for classical denotational semantics), a term representing a process, ie a state of a transition system corresponding to the activity of the process to which the statement belongs. The states of the system corresponding to the processes are given as a part of an algebraic specification (BEHAVIOUR) which formalizes a labelled transition system.

Note that the SMoLCS methodology allows to write the denotational clauses both following a continuation style (see [S]) and a direct semantics style (see [BJ]); the relationship between the two styles is studied in [AR1].

Declarations and expressions are handled analogously.

Altogether the denotational clauses can be seen as defining, inductively on the structure of the abstract syntax, a syntax-directed translation into an intermediate language for representing processes and concurrent systems.

The approach goes further. The semantics of the intermediate language is given by the algebraic specification of a <u>concurrent algebra</u> (the second step), representing a concurrent system modelling program executions. As we have seen before, the specification of the concurrent algebra consists essentially in:

-   the specification of an abstract data type formalizing a concurrent system as a labelled transition system;
-   an observational semantics associated to that abstract data type, as we have seen before.

Thus the terms of the intermediate language, obtained by the denotational clauses in the first step, can be interpreted in the concurrent algebra. In this way the denotational clauses define a homomorphism from the algebra of the abstract syntax into a semantic algebra, some carriers of which are the carriers of the corresponding sorts in the concurrent algebra.

Some advantages of this approach are:

-   in the first step it is brought to evidence what is truly sequential and what is hiddenly concurrent, thus resolving the basic interference between sequential and concurrent features;
-   the algebraic technique of the second step permits a high level of modularity and abstraction in the definition of the many structures encountered in a language; moreover it also allows to express the dependence of the semantics on some parameters formalizing the implementation dependent features;
-   the observational semantics allows to represent different semantics depending on what we want to observe of a program and to abstract from the details of the description of the transitions of the concurrent system, which is on the other hand needed, if we want to keep a close local correspondence with the usually informal operational approach of a language reference manual;
-   since the specification of the concurrent system embodies an operational idea of it as a labelled transition system, also an alternative operational approach to the semantics of the concurrent system can be taken: the axioms can be interpreted as defining labelled transition systems, to which then we can associate observational equivalences directly, a' la Milner [M1, M2].

2 SPECIFICATION OF CONCURRENT SYSTEMS

In this section we give an informal description of the SMoLCS methodology for the specification of concurrent systems together with some examples.

SMoLCS is an acronym for Structured Monitored Linear Concurrent Systems. The main idea behind SMoLCS is that, essentially, we model a concurrent system as a labelled transition system obtained by composing (structuring) the systems representing the component processes in a linear and monitored way, in a sense to be explained below.

In order to be concrete and coherent we will use as example throughout the section the specification of a concurrent system, which will be used in the following section to give the semantics of the tutorial example language EL.

2.1 *Processes as transition systems*

A set of processes is described by a labelled transition system. A labelled transition system is given by a finite presentation of three sets: states (generic element s), flags (generic element f) and transitions (triples(s,f,s') written also $s \xrightarrow{f} s'$). In the following the transitions will be defined by sets of axiom schemata of the form

$$" s \xrightarrow{f} s' \text{ IF cond }"$$

to be interpreted: if the condition cond (a boolean expression) is true, then the transition $s \xrightarrow{f} s'$ belongs to the system. It is important to note that a transition has the following intuitive meaning: a process in a state s has the capability of moving to a state s' by an action whose interaction with the external environment is represented by the flag f; hence f is conveying both information on the conditions of the environment which allow the capability to become effective and on the transformation on the environment produced by the execution of that action. This meaning of labelled transitions has now become classical after its use in CCS, [M1, M2], and in SOS, [P].

We will now illustrate it by few examples. A capability of reading the content of a cell of a shared storage by a process b can be written as

$$b \xrightarrow{READ(l,v)} b'$$

where l is a location of the shared storage and v a value.

Note that, as it will be defined in what follows, this capability will become effective only in an external environment where the content of the cell l is exactly the value v .

A capability of writing on a cell of a shared storage by a process b can be written as

$$b \xrightarrow{WRITE(l,v)} b'$$

where l is a location of the shared storage and v a value.

Analogously we can express the well known capabilities of handshake communication

$$b \xrightarrow{SEND(cid,v)} b' \text{ (sending)} \quad \text{and} \quad b \xrightarrow{REC(cid,v)} b' \text{ (receiving)}$$

where cid is a channel identifier and v a value.

An example of conditional rule is

$$b_1 ; b_2 \xrightarrow{f} b_2 \text{ IF } b_1 \xrightarrow{f} skip$$

which defines, inductively, the capabilities of a process executing the concatenation of two statements.

Example.

We now specify a labelled transition system BEHAVIOUR-SYSTEM using the metalanguage associated to SMoLCS. The states are objects, called behaviour, of sorts bh in a specification BEHAVIOUR; the flags are objects of sort act in a specification ACT, to denote kind of action; the transition relation is an operation ——> of functionality bh × act × bh → bool and the transitions are given either by conditional axioms or by equations expressing the identity of two behaviours and hence permitting to deduce the transitions of one from those of the other.

In such a definition and in the following we will use objects of sorts funct(srt_1,srt_2) for some sorts srt_1 and srt_2, these are objects defined by an algebraic specification, which behave exactly like functions with arguments of sort srt_1 and values of sort srt_2. Given two algebraic specifications A and B with main sorts a and b, A → B is a metalanguage construct for indicating the specification of the functions from elements of sort a into elements of sort b. The function declaration is expressed by the operator $\lambda\square.\square$ abstraction and the application by the operation $\square(\square)$. In the following we define the three components of BEHAVIOUR-SYSTEM by the constructs:

-   **flags** ... **end flags**, which defines the specification (named ACT) of the flags of the transition system;
-   **behaviour**... **end behaviour**, which defines the specification of the states (named BEHAVIOUR) of the transition system;
-   **actions** ... **end actions**, which defines what are the transitions of the system.

Here we intermix the formal definition with some comments (*written in italic*).

VAL *(values)*, LOC *(location of shared memory)*, CID *(channel identifiers)*, EL *(exit labels) are algebraic specifications, not further specified here, only remember that* VAL *includes boolean values.*

behaviour system
    flags
        enrich VAL + LOC + CID + BEHAVIOUR **by**
            sorts act

opns	1 TAU:	→ act
	2 ALLOC: loc	→ act
	3 READ,WRITE: loc × val	→ act
	4 SEND,REC: cid × val	→ act
	5 CREATE,BEING-CREATED: bh	→ act

    end flags

(1)   Behaviour internal action.

(2)   Allocation of an unused shared memory cell.

(3)   Reading and writing a shared memory cell.

(4)   Handshaking communication actions throughout channels.

(5)   Creating new behaviours.

HANDLER = (EL $\underset{m}{\rightarrow}$ BEHAVIOUR)[handler/map(el,bh)]

HANDLER *is the algebraic specification of the exits handlers.*

Given two specifications A *and* B *the metalanguage construct* A $\underset{m}{\rightarrow}$ B *indicates the specification of the finite maps from elements of* A *into elements of* B; A[srt$_1$/srt$_2$], *where* srt$_2$ *is a sort of* A, *indicates the specification* A *with the sort* srt$_2$ *renamed by* srt$_1$.

ENV = (VID $\underset{m}{\rightarrow}$ LOC)[environment/map(vid,loc)]

ENV *is the algebraic specification of variable environment.*

behaviour
    enrich ACT + (VAL $\rightarrow$ BEHAVIOUR) + HANDLER +
        (LOC $\rightarrow$ BEHAVIOUR) + (ENV $\rightarrow$ BEHAVIOUR) **by**
    sorts bh
    opns

	1	**nil**:	$\rightarrow$ bh
	2	**skip**:	$\rightarrow$ bh
	3	$\square$ **;** $\square$: bh $\times$ bh	$\rightarrow$ bh (assoc.)
	4	$\square \triangle \square$: act $\times$ bh	$\rightarrow$ bh
	5	**if** $\square$ **then** $\square$ **else** $\square$: val $\times$ bh $\times$ bh	$\rightarrow$ bh
{	6	**def**$_{ARG}$ $\square$ **in** $\square$: bh $\times$ funct(arg,bh)	$\rightarrow$ bh,
	7	**return**$_{ARG}$ $\square$: arg	$\rightarrow$ bh $\mid$ ARG = VAL,ENV }
	8	**cycle** $\square$: bh	$\rightarrow$ bh
	9	**trap** $\square$ **in** $\square$: handler $\times$ bh	$\rightarrow$ bh
	10	**exit** $\square$: el	$\rightarrow$ bh
	11	**choose** $\square$ **or** $\square$: bh $\times$ bh	$\rightarrow$ bh (comm., assoc.)
{	12	**choose**$_{ARG}$ $\square$: funct(arg,bh)	$\rightarrow$ bh $\mid$ ARG = LOC,VAL }

    end behaviour

(1) The null behaviour.

(2) The behaviour identity for the sequential composition operator **;** .

(3) Sequential composition of behaviours, b_1 ; b_2 *behaves as* b_1 *until it terminates then it behaves as* b_2.

(4) Prefixing of an action to a behaviour.

(5) Conditional operator (note that the elements of sort val *include also boolean values).*

(6,7) Composition of behaviours with value passing. **def**$_{VAL}$ b **in** vfunct *behaves as* b *until* b *returns a value* x, *then it behaves as* vfunct(x). *For every value* v_0 **return** v_0 *is the behaviour returning the value* v_0.

(8) The behaviour **cycle** b *executes* b *and then restarts itself.*

(9,10) Trap-exit mechanism for behaviours. The behaviour **trap** hnd **in** b *behaves as* b *until* b *performs an exit to a label* el; *if* el *is trapped by the handler* hnd, *then it behaves as specified by the handler otherwise the exit is propagated. The behaviour* **exit** el$_0$ *performs an exit to the label* el$_0$.

(11) Nondeterministic binary choice.

(12) Nondeterministic choice between infinite alternatives.

Note that BEHAVIOUR *is an example of recursive specification. The use of recursive specifications is quite natural, when trying to give specifications in a modular way. For example in this case we have that for every behaviour* b *the behaviour actions include* CREATE(b) *and that for every action* a *and behviour* b a Δ b *is a new behaviour.*

Since we use a recursive definition, we expect that the resulting specification (signature and axioms) is in some sense, a well determinated fixpoint of the transformation associated to the definition. It is not difficult to see that, under some natural conditions, such fixpoint exists. For a more technical discussion see [AR3].

actions

1 **nil** $\xrightarrow{\text{BEING-CREATED(b)}}$ b

2 a Δ b $\xrightarrow{a}$ b

3 $$\frac{b_1 \xrightarrow{a} b_1'}{b_1 \, ; \, b_2 \xrightarrow{a} b_1' ; b_2}$$

4 **skip** ; b = b   5 **exit** el ; b = **exit** el

6 **if** <u>true</u> **then** b_1 **else** b_2 = b_1 7 **if** <u>false</u> **then** b_1 **else** b_2 = b_2

{ 8 **def**$_{ARG}$ **return**$_{ARG}$ arg **in** argfunct = argfunct(arg)

9 $$\frac{b \xrightarrow{a} b'}{\text{def}_{ARG} \ b \ \text{in argfunct} \ \xrightarrow{a} \ \text{def}_{ARG} \ b' \ \text{in argfunct}}$$ | ARG = VAL,ENV }

10 **trap** hnd **in skip** = **skip**

11 **trap** hnd **in exit** el = hnd(el) **IF** el ∈ <u>dom</u> hnd = <u>true</u>

12 **trap** hnd **in exit** el = **exit** el **IF** el ∈ <u>dom</u> hnd = <u>false</u>

13 $$\frac{b \xrightarrow{a} b'}{\text{trap hnd in b} \ \xrightarrow{a} \ \text{trap hnd in b'}}$$

14 **cycle** b = b ; **cycle** b   15 **cycle exit** el = **exit** el

16 $$\frac{b_1 \xrightarrow{a} b_1'}{\text{choose} \ b_1 \ \text{or} \ b_2 \ \xrightarrow{a} b_1'}$$

{ 17 $$\frac{\text{argfunct(arg)} \xrightarrow{a} b}{\text{choose}_{ARG} \ \text{argfunct} \xrightarrow{a} b}$$ | ARG = LOC,VAL }

end actions

end behaviour system *End example.*

2.2 *Concurrent Systems*

In a concurrent system a state consists of the states of the component processes plus some global information; we choose to represent it as a couple $(s_1| s_2|...| s_n,i)$ where i is the global information and $s_1|s_2| ...|s_n$ is a multiset of states of the components; the symbol | suggests that the states are "in parallel" and can be formally defined as the union on multisets, after identifying a singleton multiset with its unique element.

Now, assuming that a basic transition system BTS, representing the component processes, is given, how do we specify the resulting composed system?

Our idea is to split the composition in some steps. First the actions of the basic system are composed producing new actions; this step is conveniently subdivided in two other steps: one (synchronization) defines the actions resulting from some synchronized cooperation between processes; another (parallel composition) defines which are the synchronous actions that can happen in parallel. Then a third step (monitoring) defines which actions resulting after the second step are allowed to happen as actions of the whole system; the selection of the actions may depend on some appropriate information.

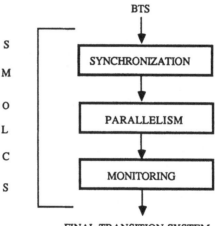

FINAL TRANSITION SYSTEM

Before detailing the three steps, let us explain what does linearity mean for us. Every action of the composed system is a linear composition of some actions of the basic system in the following sense:
if $s_1 \xrightarrow{f_1} s_1'$ and $s_2 \xrightarrow{f_2} s_2'$ are two actions, then an action $s_1|s_2 \xrightarrow{sf} s_1'|s_2'$ for some sf, is a linear composition of them.

Now every action of the composed system, say $(ms,inf) \xrightarrow{sf} (ms',inf')$, will be of the form $(ms_1|ms_2,inf) \xrightarrow{sf} (ms_1'|ms_2,inf')$, where for some sf' $(ms_1,inf) \xrightarrow{sf'} (ms_1',inf')$ is a linear composition of some basic actions. Thus every system action is a linear action.

We are now going to make the previous concepts a bit more precise.

Example.

As an example of concurrent system we report the definition written in a somewhat friendly syntax of CONC_SYST, which is needed for giving the semantics of the example language EL. Following the above schema the definition is split in four subconstructs which are reported and commented in sections: 2.1 (the basic system), 2.2.1 (synchronization), 2.2.2 (parallelism), 2.2.3 (monitoring).

End example.

2.2.1 Synchronization

We define the synchronous actions by giving a new transition system STS, where the transition relation ⟶ corresponds to the synchronous actions themselves, starting from a basic system BTS (representing the component processes with transition relation —>). The states of STS are couples (ms,i) where ms is a multiset of states of the basic system and i some global (with respect to the component processes) information.

The synchronous actions are given by the following axiom schema

$$(s_1|...|s_n,i) \xrightarrow{sf} (s_1'|...|s_n',i') \quad \text{IF} \quad (\bigwedge_{1 \le j \le n} s_j \xrightarrow{f_j} s_j') \wedge$$

$$Iss(sf,f_1| ... |f_n,i) = \underline{true} \wedge Sit(sf,i,i') = \underline{true}$$

where Iss (for <u>Is</u> a <u>s</u>ynchronous action) and Sit (for <u>S</u>ynchronous <u>i</u>nformation <u>t</u>ransformation) are two partial functions, whose meaning is the following:

if $Iss(sf,f_1| ... |f_n,i) = \underline{true}$ then the basic actions with flags $f_1,...,f_n$ when the global information part is i may synchronize producing an action with flag sf;

if $Sit(sf,i,i') = \underline{true}$ then the execution of the action with flag sf changes the global information i into i'.

Note that the transformation of the global information associated to a synchronous action can be nondeterministic.

We require that whenever $Iss(sf,f_1|...|f_n,i)$ is equal to <u>true</u> then there exists an i' such that $Sit(sf,i,i')$ is equal to <u>true</u>.

Iss and Sit are defined by means of a set of conditional axioms.

We briefly illustrate the idea referring to the example capabilities of subsection 2.1.

The effect of reading a location of the shared storage can be defined by:

    $Iss(READ(l,v),READ(l,v),i) = \underline{true} \quad \text{IF } i(l) = v$

    $Sit(READ(l,v),i,i) = \underline{true}$

where in the global information i is recorded the state of the shared storage; note that the flag of the resulting action is still READ(l,v) (thus recording the kind of the action) since the effective happening of the action will still depend on the actions of the other processes because two contemporaneous readings of the same locations are mutually exclusive; we will handle that in the parallel composition step. Note that, for clarity, we use a different set of characters for writing the flags of synchronous actions, even when they coincide with the flags of the basic transition system.

The effect of writing on a location of the shared storage is defined by:

    $Iss(WRITE(l,v),WRITE(l,v),i) = \underline{true}$

Sit(**WRITE**(l,v),i,i[v/l])) = <u>true</u> ;

i[v/l] represents the state of the storage memory where the content of the location l has been changed in v .
The effect of a handshaking communication is defined by

Iss(**TAU**,SEND(cid,v) | REC(cid,v),i) = <u>true</u>   Sit(**TAU**,i,i) = <u>true</u>

where the synchronous flag **TAU** reminds Milner's symbol for an internal action, ie an action with no interaction with the external environment. Moreover no other synchronous actions involving SEND(cid,v) or REC(cid,v) are defined, thus a SEND(cid,v) action of a behaviour can be executed only together with a REC(cid,v) action of another behaviour.

Note that also the process internal actions will become synchronous actions, as defined below

Iss(**TAU**,TAU,i) = <u>true</u>.

Creation and termination of component processes are handled by defining a particular process state **nil** with the property **nil**$|b_1|...|b_n = b_1|...|b_n$, process actions such as:

nil $\xrightarrow{\underline{\text{BEING-CREATED(b)}}}$ b, b $\xrightarrow{\underline{\text{CREATE(b_1)}}}$ b_1,

b $\xrightarrow{\underline{\text{END}}}$ **nil**

and synchronous actions such as

Iss(**TAU**,BEING-CREATED(b)|CREATE(b),i) = <u>true</u>,

Iss(**TAU**,END,i) = <u>true</u>.

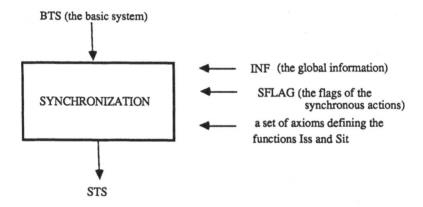

From what said before for defining a synchronization operation on a basic transition system it is sufficient to give the global information, the flags of the synchronous actions and the definitions of the functions Iss and Sit.

Example.

As example of synchronization step we report the definition written in a somewhat friendly syntax of the synchronization used in the definition of the concurrent system CONC_SYST, needed for giving the semantics of the example language EL.

The definition is split in three constructs:

- **global information** ... **end global information**, which defines the algebraic specification (named INF) of the global information;

- **synchronous flags** ... **end synchronous flags**, which defines algebraic the specification (named SFLAG) of the synchronous action flags;

- **synchronous actions** ... **end synchronous actions**, which defines what are the synchronous actions.

We intermix the formal definition with some comments (*written in italic*).

synchronization
    global information
        enrich (LOC $\rightleftharpoons$ VAL)[inf/map(loc,val)] **by**
            opns Freeloc: inf → loc
            axioms (Freeloc(i) ∈ **dom** i) = **false**
    end global information

The global information, in this case, consists of the state of the shared storage abstractly specified as a map from locations into values enriched by an operation Freeloc, which, taken a storage state, returns a location unused in that state.

    synchronous flags   ACT[sflag/act]   **end synchronous flags**

The flags of the synchronous actions are just the same as the flags of the actions of the basic system.
To improve readability we usually use a different set of characters to write synchronous flags.

synchronous actions

(1)
$$\frac{b \xrightarrow{\text{TAU}} b'}{b \xrightarrow{\text{TAU}} b'}$$
nocond
notransf

(2)
$$\frac{b \xrightarrow{\text{ALLOC(l)}} b'}{b \xrightarrow{\text{WRITE(l,Undef)}} b'}$$
cond: Freeloc(i) = l
transf: i + [l → Undef]

(3)
$$\frac{b \xrightarrow{\text{READ(l,v)}} b'}{b \xrightarrow{\text{TAU}} b'}$$
cond: i(l) = v
notransf

(4)
$$\frac{b \xrightarrow{\text{WRITE(l,v)}} b'}{b \xrightarrow{\text{WRITE(l,v)}} b'}$$
nocond
transf: i + [l → v]

(5)
$$\frac{b_1 \xrightarrow{\text{SEND(cid,v)}} b_1' \wedge b_2 \xrightarrow{\text{REC(cid,v)}} b_2'}{b_1|b_2 \xrightarrow{\text{TAU}} b_1'|b_2'}$$
nocond
notransf

$$(6) \quad \frac{b \xrightarrow{\text{CREATE}(b_1)} b' \ \wedge \ \text{nil} \xrightarrow{\text{BEING-CREATED}(b_1)} b_1}{b|\text{nil} \xrightarrow[\substack{\text{nocond} \\ \text{notransf}}]{\text{TAU}} b'|b_1}$$

end synchronous actions

First note that the rules of synchronous actions are given in a friendly syntax which makes explicit the synchronized transitions instead of defining just the functions Iss and Sit as in the basic synchronization schema explained above, which instead emphasizes the underlying mathematics. But clearly this friendly version is nothing more than a sugared version for readability purposes.

Synchronous actions are defined by means of a syntactic construct, called rule, because its form reminds inference rules.

In every rule the actions of the basic system to be synchronized are reported above the line: for example the actions $b_1 \xrightarrow{\text{SEND}(cid,v)} b_1'$ and $b_2 \xrightarrow{\text{REC}(cid,v)} b_2'$ in rule (5) (defining the handshaking communication) or a single internal behaviour action $b \xrightarrow{\text{TAU}} b'$ as in rule (1).

The flag of the synchronous action defined by a rule is written above the arrow ($\longrightarrow$) under the line: for example rules (1), (3), (5) and (6) define actions flagged by **TAU** *(this flag means that these actions do not need further considerations in the following composition steps (parallelism and monitoring)). Instead rules (2) and (4) define actions flagged by* **WRITE**(l,v) *to record that these actions use the location l of the shared storage; this information will be used in the parallel composition step.*

The boolean term after **cond**: *indicates the condition on the global information under which the action can be performed; for example in rule (3) (reading the contents of the location l) the condition is that the actual contents of the location l is just v.*

nocond *is used when the defined action is not conditionated on the global information.*

The term of sort inf *following* **transf**: *indicates the transformation of the global information due to the synchronous action defined by the rule. For example rule (4) (writing on the shared storage) states that the contents of the location l of the shared storage has to be changed to v.* **notransf** *is used when the defined action does not transform the global information.*

End example.

2.2.2 Parallelism

Intuitively by means of this operation we define whether two actions can be executed in parallel (without synchronization). The actions to be considered for composition are, inductively, the actions of the synchronized system STS and the new actions already obtained by parallel composition.

As before for synchronization we can describe the operation of parallel composition as producing a new system PTS from the system STS. PTS is simply given by augmenting the transitions of STS (indicated by $\rightarrow$) with the new elements, which are given by the following axiom schema:

$$(ms_1|ms_2,i) \xrightarrow{sf_1 /\!\!/ sf_2} (ms_1'|ms_2',i')$$

IF $(sf_1 /\!\!/ sf_2$ is defined$) \wedge (ms_1,i) \xrightarrow{sf_1} (ms_1',i_1') \wedge$

$(ms_2,i) \xrightarrow{sf_2} (ms_2',i_2') \wedge Pit(sf_1 /\!\!/ sf_2,i,i') = \underline{true}$

provided we have given the partial (binary, commutative and associative) operation $/\!/$ on sflags and the partial function Pit (for Parallel action information transformation) which defines the transformation (in general nondeterministic) of the information due to the composed action with flag $sf_1 /\!/ sf_2$. Moreover we require that whenever

$$(sf_1 /\!/ sf_2 \text{ is defined}) \wedge (ms_1,i)\xrightarrow{sf_1} (ms_1',i_1') \wedge (ms_2,i) \xrightarrow{sf_2} (ms_2',i_2')$$

holds then there exists i' such that $Pit(sf_1 /\!/ sf_2,i,i')$ is equal to <u>true</u> .

In our previous examples a writing action on the shared storage and a handshaking communication can be executed together giving a new composed action, which in turn can be executed together with handshaking action of some other processes. On the converse a reading action of a cell of the shared storage does exclude whatever other action on the same storage cell; thus we have

TAU $/\!/$ **sf** is defined for every flag sf

READ(l,v) $/\!/$ **READ**(l,v') is not defined for any value v and v'

READ(l,v) $/\!/$ **WRITE**(l,v') is not defined for any value v and v'.

For giving a parallel composition thus it is sufficient to give the axioms defining the operations $/\!/$ and Pit.

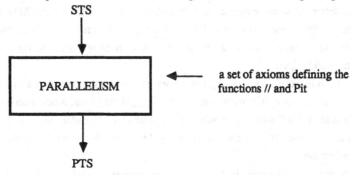

Example.

As an example of the parallelism step we report the definition (written in the FD metalanguage) of the parallelism used in the definition of the concurrent system CONC_SYST, needed for giving the semantics of EL.

(1) **D(TAU//sf)**

(2) **D(WRITE(l1,v1) //WRITE(l2,v2))** IF (l1=l2) $=$ <u>false</u>

(3) **D(READ(l1,v1) //WRITE(l2,v2))** IF (l1=l2) $=$ <u>false</u>

(4) **D(READ(l1,v1) //READ(l2,v2))** IF (l1=l2) $=$ <u>false</u>

(5) **D((sf1//sf2)//sf3)** IF **D(sf1//sf3)**$\wedge$ **D(sf2//sf3)**

(6)   Pit(TAU,i,i) $=$ <u>true</u>   Pit(READ(l,v),i,i) $=$ <u>true</u>   Pit(WRITE(l,v),i,i+[l$\rightarrow$v]) $=$ <u>true</u>

(7) Pit(sf1//sf2,i,i") IF Pit(sf1,i,i') $=$ <u>true</u> $\wedge$ Pit(sf2,i',i") $=$ <u>true</u>

The first four axioms state that two contemporaneously writings and/or readings of the same location of the shared storage are not allowed and that these are the only mutually exclusive actions.

Note that the transformation of the information due to a composed action is given by the composition of the transformations due to the component actions (axiom 7). Note also that for all the composable couples of actions the order in which the transformations are composed does not matter.
End example.

2.2.3 Monitoring

Here we take into consideration any form of global control, by which only some of the actions which are locally possible in a system (ie those obtained by (synchronization and) parallel composition) are allowed to become actions of the overall system. It is at this step that we can, for example, define an interleaving mode, admitting only one synchronized action at time, or a mode in which all actions that can be executed together do so . Here we can also define that the buffer reading actions take precedence on the buffer writing actions (ie when in a state there is a reading action, a writing action will never be allowed). We also assume that the control may depend on some ad hoc information (monitor information), that can be updated; by this we can specify, eg a scheduling based on process priorities.

As before we can define the monitoring operation by giving a new transition system MTS (with transition relation ⟶) starting from a parallel system PTS (with transition relation —>). The states of MTS will be couples (sst,mi) where sst is a state of PTS and mi some monitor information. The transitions of the new system are defined by giving:

- a partial boolean function Mon (for $\underline{Mon}$itoring) which, taken a state (ms,i) and a flag of the parallel system (sf) and a monitor information (mi), returns <u>true</u> if the partial action with flag sf is allowed in the system state (ms,i) when the monitor information is mi. The other argument of Mon (extf) is the flag of the resulting action of the new system and indicates the interaction of that action with the external environment;

- a function Mit (for $\underline{M}$onitor $\underline{i}$nformation $\underline{t}$ransformation) which gives the (in general nondeterministic) transformation of the monitor information associated to the execution of the allowed parallel actions (Mit(sf,mi,mi') = <u>true</u> iff mi' is a possible resulting monitor information after the execution of the parallel action sf).

Then an action of the system can be specified by the following axiom schema:

$$((ms_1|ms_2,i),mi) ==extf==> ((ms_1'|ms_2,i'),mi')$$

$$\text{IF} \quad (ms_1,i)\xrightarrow{sf} (ms_1',i') \ \wedge \ Mon(sf,(ms_1|ms_2,i),mi,extf) = \underline{true} \ \wedge$$
$$Mit(sf,mi,mi') = \underline{true}.$$

Note that the axiom specifies, as it was anticipated informally, that an action of the system is determined by an action of a part of the component processes; here the partial action is

$(ms_1,i) \xrightarrow{sf} (ms_1',i')$ and ms_2 is the multiset of the states that do not cooperate to that action. Moreover, since every parallel action is the linear composition of some basic actions, the axiom also says that every system action is a linear action.

For defining a monitoring operation it is sufficient to give the monitor information, the flags of the resulting actions (external flags) and the definitions of the functions Mon and Mit.

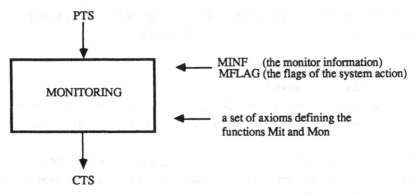

PTS

MONITORING

MINF   (the monitor information)
MFLAG (the flags of the system action)

a set of axioms defining the
functions Mit and Mon

CTS

Example.

As an example of monitoring we report the definition (written in the metalanguage associated to our methodology) of the monitoring used in the definition of the concurrent system CONC_SYST, needed for giving the semantics of the example language EL.

The definition is in three constructs:

- **monitor information** ...**end monitor information** which defines the algebraic specification (named MINF) of the monitor information;
- **external flags** ... **end external flags** which defines the algebraic specification (named EXT-FLAG) of the external flags;
- **monitoring axioms** ... **end monitoring axioms** which defines the operations Mon and Mit.

monitor information   NULL[minf/null, Null_Minf/Null]   **end monitor information**

external flags   NULL[extflag/null, TAU/Null]   **end external flags**

monitoring axioms

$$(ms_1,i)\xrightarrow{\quad sf \quad}(ms_1',i')$$

$$((ms_1|ms,i),mi) ===TAU===>((ms_1'|ms,i'),mi)$$
cond: true

end monitoring axioms

NULL *indicates the specification* **sort** null **opns** Null → null.
In this case the specifications of the monitor information and of the external flags have just only one element because the monitoring is constant (ie does not depend on some varying information) and the system is closed (ie the system does not interact with some external world).
In our example we have a free monitoring, ie every parallel action is allowed to be performed, and

obviously the monitor information is never changed. We have chosen this particular form of monitoring because we have no assumptions on the duration of the behaviour actions.

End example.

2.3 Semantics of transition systems

After having defined a transition system we must fix which is the semantics to which we are interested; ie define a semantic equivalence on the system states, identifying all the states we are not interested to distinguish.

The most simple case is when we want to define an operational semantics ie a semantics such that two states of the system are semantically equivalent iff they have the same associated derivation tree, where all intermediate states and flags are observable:

let s_1, s_2 be two states of the system

s_1 operationally equivalent to s_2 **iff**   for every s,f state and flag of the system
$$s_1 \xrightarrow{f} s \text{ iff } s_2 \xrightarrow{f} s .$$

But more general as we want to be able to give different semantics to a transition system depending on what we want to observe of the system; that seems to be essential in concurrency where now it is recognized that no single semantic equivalence can capture the variety of meanings, while preserving a sufficient level of abstraction. Someone, for example, may be interested in observing the input-output relation of a program (recall that SMoLCS could be used also to give the semantics of concurrent programming languages, see section 3) or in observing the communications with the external environment which the system is able to perform. In what follows we will make precise the idea of observational semantics and give some examples.

To give an observational semantics to a transition system it is sufficient to define some boolean observation functions which associate to the system states some observation values; then two states are semantically equivalent iff they have associated the same observation values.

Let STATE be the set of the states of the transition system TS,

$$\text{obsfunct}_1 : \text{STATE} \times \text{OBS}_1 \rightarrow \text{BOOL}$$
....
$$\text{obsfunct}_r : \text{STATE} \times \text{OBS}_r \rightarrow \text{BOOL}$$

be the observation functions and

$\text{OBS}_1, ..., \text{OBS}_r$ be the sets of observation values.

For every $s_1, s_2 \in \text{STATE}$ s_1 observationally equivalent to s_2

iff

for every j $1 \le j \le r$, for every $o_j \in \text{OBS}_j$
$$\text{obsfunct}_j(s_1, o_j) = \text{obsfunct}_j(s_2, o_j) .$$

Example.

As an example of semantic equivalence on a transition system we report the definition of the semantic equivalence which we impose on the concurrent system CONC_SYST, used for giving the semantics of the example language EL.

observational semantics

    observations   INF **end observations**

    observations operations

        Is_Result : state $\times$ inf $\rightarrow$ bool

        Is_Result(((**skip**,i),mi),i) = <u>true</u>

        Is_Result(s,i) = <u>true</u> IF s $\xrightarrow{\text{TAU}}$ s' $\wedge$ Is_Result(s',i) = <u>true</u>

    end observation operations

In this case we are interested in an input/output semantics on the concurrent systems where the outputs are just the final states of the shared storage (the global information parts). For simplicity here we do not care about nontermination: two states of the concurrent system are equivalent iff they produce the same finite results.

Being CONC_SYST defined using the SMoLCS metalanguage and then algebraically as an algebraic specification, the above observational semantics permits to define a Sig(CONC_SYST)-algebra, called CALG (see for a full formal explanation [ARW]) such that:

for every s_1, s_2 terms of sort state

    CALG $\models s_1 = s_2$ iff s_1 observationally equivalent to s_2 ;

moreover for every t_1, t_2 terms of sort srt (srt $\in$ Sorts(CONC_SYST))

    CALG $\models t_1 = t_2$ iff for every context s[ ] of sort state with a hole of sort srt

                    s[t_1] is observationally equivalent to s[t_2].

End example.

2.4 Hierarchical concurrent systems

In the previous sections we have defined a three step procedure that, given a basic transition system, specifying some component processes, and a synchronization, a parallel composition and a monitoring, produces a new transition system, specifying a concurrent system. Clearly the procedure can be iterated; if the component processes are themselves concurrent systems, then they can be specified by the same procedure. Consider for example a net of (workstations) nodes, such that in a node many processes can cooperate, possibly using a shared memory, while the nodes can exchange messages in a broadcasting or/and point to point mode. Then we can specify the net applying twice the SMoLCS procedure; in one application the basic transition system consists of the specification of the processes cooperating in a node and the resulting concurrent system specifies a node; in the other one the nodes, specified in the first level, become the new basic transition system and the resulting concurrent system specifies the net. It can also be shown that the procedure can be applied inductively; hence it is possible to specify systems in which a component process has the same nature of the composed processes, as in CCS, ie where states and transitions of the final system are embedded into the states and the transitions of the basic system.

3 DENOTATIONAL CLAUSES

Here we illustrate the step of our approach connecting the abstract syntax of a language to a semantics in a truly denotational style. The general idea is that we associate to a well formed construct a term on the signature of the concurrent algebra. This step can be seen as a translation into some algebra of words. But considering the interpretation of the terms in the concurrent algebra, we get a value as in the usual denotational style.

In order to explain the main concepts, we define a toy example language and explain the related denotational clauses. We use a direct semantics style, with VDM-like constructs; the standard reference for the concepts of denotational semantics and notations used are [S] and [BJ], of course for sequential languages.

Recall that the concurrent algebra associated to the example is the one defined in section 2, as example of SMoLCS specifications. In what follows that concurrent algebra will be indicated by CALG.

See [AR] for a treatment using the continuation semantics style and [AR1] for a formal comparison with equivalence proofs of the two styles.

3.1 *The example language EL*

We give the abstract syntax of EL by giving an algebraic specification AS using the metalanguage associated to the methodology and then choosing its initial model; the various constructs will be shortly commented.

<u>spec</u> AS =
    <u>enrich</u> EID + VID + CID <u>by</u>
        <u>sorts</u> program, stat, handler-part, decs, exp
        <u>opns</u>

1	<u>program</u> □: stat	→ program	(total)
2	Λ:	→ handler-part	(total)
3	□ → □, □: eid × stat × handler-part	→ handler-part	(total)
4	Λ:	→ decs	(total)
5	□;□: vid × decs	→ decs	(total)
6	□ := □: vid × exp	→ stat	(total)
7	□;□: stat × stat	→ stat	(total)
8	if □ <u>then</u> □ <u>else</u> □: exp × stat × stat	→ stat	(total)
9	<u>while</u> □ <u>do</u> □: exp × stat	→ stat	(total)
10	<u>raise</u> □: eid	→ stat	(total)
11	□ <u>begin</u> □ □ <u>end</u>: decs × stat × handler-part	→ stat	(total)
12	<u>send</u>: cid × exp	→ stat	(total)
13	<u>rec</u>: cid × vid	→ stat	(total)
14	<u>create-proc</u> □: stat	→ stat	(total)
15	<u>choose</u> □ <u>or</u> □: stat × stat	→ stat	(total)
16	{c_i:	→ exp| i ≥ 0 }	-- constants
17	{□ bop_i □: exp × exp	→ exp | i ≥ 0}	
			--binary operators
18	E: vid	→ exp	

Usually we write simply vi instead of E(vi).

Identifiers of exceptions (EID), variables (VID) and channels (CID) are not further specified here.

<u>Comments</u>. A program is just a statement (1). Together with some usual sequential statements (6,...,9),

there is a mechanism for raising (10) and handling (3) exceptions: if eid → st appears in the handler part of a block, then when the exception eid is raised, the block execution is abandoned and the statement st is executed in substitution.

The concurrent structure is generated by the creation of processes (14) which are statements; all processes can be executed in parallel and there are no constraints on the duration of their actions.

The variables declared in a block (5) are shared by the process executing the block itself and by the processes created within the block. Note that also the program can be considered a process. Together with shared variables, there is a provision for handshaking communication via channels between processes (12,13) and for nondeterministic choice (15).

Clearly, due to the presence of shared variables, the execution of usual sequential constructs involves concurrent interactions among processes. Note how the concurrent structure is rather hidden by the syntactic structure, which is typical of a block structured language.

>   **type** PROGRAM = AS

The metalanguage construct for primitive type declarations **type** PTID = SPEC-EXPR declares the primitive type PTID whose elements are the the elements of the carrier of the main sort of a particular model (an algebra) of the algebraic specification SPEC-EXPR and also for every sort srt of SPEC-EXPR declares a primitive type SRT whose elements are the elements of the carrier of srt in that model. That particular model of SPEC-EXPR is chosen as follows:

- if SPEC-EXPR is the specification of the concurrent system (CONC_SYST) defined using the SMoLCS methodology or one of its subspecifications the chosen model is the concurrent algebra or the corresponding subalgebra;
- otherwise it is the initial model of SPEC-EXPR, indicated by $I_{SPEC\text{-}EXPR}$.

Thus, since AS is not a subalgebra of CONC_SYST

>   PROGRAM = $(I_{AS})_{program}$
>   STAT = $(I_{AS})_{stat}$ and so on

(if A is an algebra and srt is a sort of A, then A_{srt} indicates the carrier of srt in A).

3.2 Semantic domains and functions

>   **type** ANSWER = STATE

STATE with main sort state is a subspecification of CONC_SYST, thus

>   ANSWER = $CALG_{state}$;

note that the specification STATE defines the states of the transition system represented by CONC_SYST.

>   **type** PVAL = BEHAVIOUR

BEHAVIOUR with main sort bh is a subspecification of CONC_SYST thus

>   PVAL = $CALG_{bh}$;

moreover BEHAVIOUR has a sort environment and a sort handler, thus we have also the types

>   ENVIRONMENT = $CALG_{environment}$,

HANDLER = CALG$_{handler}$ ·

The semantic functions are:

P : PROGRAM → ANSWER

S : STAT → ENVIRONMENT → PVAL

E : EXP → ENVIRONMENT → PVAL

D : DECS → ENVIRONMENT → PVAL

H : HANDLER-PART → ENVIRONMENT → HANDLER.

In writing the clauses we will use the following shortcuts:

def$_{ARG}$ b in λ v . b' is simply written def v = b in b', return$_{ARG}$ v is simply written return v and choose$_{ARG}$ λ v . b is simply written choose v: ARG in b.

1   P[program st] ≙ ((S[st]env$_0$,stg$_0$),Null_Minf)

where env$_0$ is the initial empty environment, stg$_0$ is the initial state of the global shared storage and Null_Minf the null monitor information.

2   S[x := e]env ≙ def v = E[e]env in WRITE(env(x),v) Δ skip

where skip: → bh is an operator on behaviours defined in section 2.

3   S[st$_1$; st$_2$]env ≙ S[st$_1$]env ; S[st$_2$]env

where □ ; □: bh × bh → bh (assoc) is an operation on behaviours defined in section 2.

4   S[if be then st$_1$ else st$_2$]env ≙ def v = E[be]env in if v then S[st$_1$]env else S[st$_2$]env

5   S[while be do st]env ≙ trap [End-While → skip] in

                               cycle def v = E[be]env in if v then S[st]env else exit End-While

6   S[decs begin st hnd end]env ≙ def env' = D[decs]env in

                               trap H[hnd]env' in$S$[st]env'

7   S[raise eid]env ≙ TAU Δ exit eid

8   S[send(cid,e)]env ≙ def v = E[e]env in SEND(cid,v) Δ skip

9   S[rec(cid,x)]env ≙ choose v: VAL in REC(cid,v) Δ WRITE(env(x),v) Δ skip

10   S[create-proc st]env ≙ CREATE(S[st]env) Δ skip

11   S[choose st$_1$ or st$_2$)]env ≙ choose S[st$_1$]env or S[st$_2$]env

12    $D[\Lambda]$env $\triangleq$ **return** env

13    $D[x; decs]$env $\triangleq$ **choose** l: LOC **in** ALLOC(l) Δ $D[decs](env+[x \rightarrow l])$

14    $E[c_i]$env $\triangleq$ TAU Δ **return** $Const[c_i]$ for every $i \geq 0$
where $Const$ is the semantic function for constants which is not further specified here.

15    $E[e_1 \ bop_i \ e_2]$env $\triangleq$ **def** $v_1 = E[e_1]$env **in**
                        def $v_2 = E[e_2]$env **in** TAU Δ **return** $Bop[bop](v_1, v_2)$ for every $i \geq 0$
where Bop is the semantic function for binary operators which is not further specified here.

16    $E[x]$env $\triangleq$ **choose** v: VAL **in** READ(env(x),v) Δ **return** v

17    $H[\Lambda] \triangleq []$
where [] denotes the empty map.

18    $H[eid \rightarrow st,hnd]$env $\triangleq$ $(H[hnd])+ [eid \rightarrow S[st]env]$.

Comments. Clause 1 defines the semantics of a program consisting of a statement st as the value in the concurrent algebra CALG of the term $((S[st]env_0, stg_0), Null\_Minf)$.
That is the syntactic representation of a state of the concurrent transition system CONC_SYST defined in section 2, whose states correspond to the execution states of the EL programs.
Formally $((S[st]env_0, stg_0), Null\_Minf)$ will be a term of sort state in a specification STATE, subspecification of CONC_SYST (the states of the concurrent system); its semantic value, as already said, is an equivalence class corresponding to an observational semantics on CONC_SYST (algebra CALG). For ease of notation and for a reason indicated in 3.3, here and in all clauses we will simply indicate by t the semantic value t^{CALG} of a term t in CALG.
Note that $S[st]env_0$ defines, inductively, a behaviour; intuitively a behaviour represents a process and formally is a term of sort bh in an algebra of behaviours BEHAVIOUR, defined as a subtype of CONC_SYST. A state of CONC_SYST has three components: the first is a multiset of behaviours, the second represents some information global to all behaviours (in the case of EL the state of the shared storage) and the third some information needed for the monitoring step (in the case of EL it is null).
In clause 2, as usual env is a generic element in ENVIRONMENT; v is a variable of type VAL. WRITE(env(x),v) Δ **skip** is a behaviour, whose only action is to write v in the location env(x) (**skip** is an identity element for the behaviour sequential composition operation); the symbol Δ has the same role as the dot in the CCS behaviour α. be.
Behaviour actions will always be indicated by strings of capital letters and hyphens.
$E[e]$env is a behaviour which evaluates the expression e and returns its value.
def x = b **in** b_1 is a behaviour which behaves as b until it returns a value v and then as $(\lambda x . b_1)(v)$.

Note that the term λ y . ... is a term built on the signature of CALG and intuitively corresponds to a function; formally it is a term of sort funct(val,bh) in a specification VAL →BEHAVIOUR (subtype of STATE) (→ is a metalanguage operator) representing the purely algebraic counterpart of a space of functions (as it was done, more or less, in [BW3], when defining algebraically the semantics of functional languages).

Clause 3 significantly look the same as in the purely sequential case, as clauses 6 and 12. Only note that □; □: bh × bh → bh is the sequential composition of processes and that the construct **trap** hnd **in** b traps exits executed by the behaviour b with the handler hnd, see a precise definition in section 2.

Note in clauses 4 and 5 that **if** □ **then** □ **else** □ is an operation on behaviours and most importantly in clause 5 how fixpoints are handled. **cycle** □: bh × bh → bh is an operation on behaviours characterized by the axiom **cycle** b = b; **cycle** b. Note that the execution of an exit allows to abbandon what follows it; then gives a way to stop recursion.

Clause 16 makes explicit some hidden concurrency related to the evaluation of a shared variable.

The idea behind the behaviour **choose** v: VAL **in** READ(env(x),v) Δ **return** v, where env(x) is a location, is that the content of the location env(x) will depend on the moment when the process block, in which the evaluation of x is performed, gets access to the shared storage; if v_0 is the value in env(x) at that moment, then the action READ(env(x),v_0) will be performed and the value v_0 will be returned.

As said before a term of form λ v. READ(l,v) Δ **return** v represents a function from values to behaviours and is a term of sort funct(val,bh) in a specification VAL →BEHAVIOUR.

The term **choose** v: VAL **in** READ(l,v) Δ **return** v represents a process which is the infinite sum, for v ranging in VAL, of the processes READ(l,v) Δ **return** v ; in Milner's notation (see [M2]) it would be

Σ READ(l,v) Δ **return** v; we prefer for technical reasons to consider **choose** as an operation of VAL

functionality funct(val,bh) → bh.

Formally, as it is defined in the specification of BEHAVIOUR-SYSTEM, clause 13 has to be interpreted analogously. Clause 7 shows a typical situation, also shared by clauses 14 and 15: it would be as in the purely sequential case, but in the concurrent case an action is performed, consisting, for example in clause 7 in transferring the control to the exception handler, whose value is hnd(eid); this action is internal, ie does not involve interactions with other processes or with the global information, and hence it is indicated by TAU following Milner's notation. Representing explicitly that internal action is not necessary whenever the final observational semantics, analysed by CALG, does care only in the results of the actions and not on their ordering.

Clauses 8, 9 and 10 are usual clauses for concurrent and nondeterministic statements; only recall the comment to clauses 16 for interpreting the **choose** operation in 9 and note that the results of the action capabilities REC, SEND will be defined in the definition of the concurrent system CONC_SYST, see section 2.

Consistency of denotational clauses can be checked in two steps: first it is proved that the semantic functions are total on statically correct programs ie they associate to every correct construct a term on the signature of the specification CONC_SYST; secondly it is checked that all the associated terms have a defined interpretation in the concurrent algebra CALG.

These facts are true of EL but we do not give here the proof. Here we show on two examples how our definition works.

An assignment statement.

Let $\underline{3}, \underline{100}$ be two constants and $\oplus, <$ be two binary operators, whose intuitive meaning is clear.

$S[x := x \oplus \underline{3}]$env =   (clause 2)

def $v = E[x \oplus \underline{3}]$env **in**
   WRITE(env(x),v) Δ **skip** =   (clause 15)

def $v = ($ **def** $v_1 = E[x]$env **in**
       def $v_2 = E[\underline{3}]$env **in**
        TAU Δ **return** $Bop[\oplus](v_1, v_2))$ **in**
   WRITE(env(x),v) Δ **skip** =   (clause 16)

def $v = ($ **def** $v_1 = ($**choose** v: VAL **in** READ(env(x),v) Δ **return** v) **in**
       def $v_2 = E[\underline{3}]$env **in**
        TAU Δ **return** $Bop[\oplus](v_1, v_2))$ **in**
   WRITE(env(x),v) Δ **skip** =   (clause 14)

def $v = ($ **def** $v_1 = ($**choose** v: VAL **in** READ(env(x),v) Δ **return** v) **in**
       def $v_2 = $ TAU Δ **return** $Const[\underline{3}]$ **in**
        TAU Δ **return** $Bop[\oplus](v_1, v_2))$ **in**
   WRITE(env(x),v) Δ **skip** =

def $v = ($ **def** $v_1 = ($**choose** v: VAL **in** READ(env(x),v) Δ **return** v) **in**
       def $v_2 = $ TAU Δ **return** 3 **in**
        TAU Δ **return** $v_1 + v_2$) **in**
   WRITE(env(x),v) Δ **skip** $= b_1$.

A while statement.

$S[\underline{\text{while}}\ x < \underline{100}\ \underline{\text{do}}\ x := x \oplus \underline{3}]$env=   (clause 5)

trap [End-While $\rightarrow$ **skip**] **in**
   cycle **def** $v = E[x < \underline{100}]$env **in**
        if v   **then** $S[x := x < \underline{100}]$env
               else **exit** End-While   = (*)

By applying clauses 15,16 and 14 we have

$E[x < \underline{100}]$env =
def $v_1 = ($**choose** v: VAL **in** READ(env(x),v) Δ **return** v) **in**
def $v_2 = $ TAU Δ **return** 100 **in**
  TAU Δ **return** $v_1 < v_2$

thus (*) =
trap [End-While $\rightarrow$ **skip**] **in**
   cycle **def** $v = ($ **def** $v_1 = ($**choose** v: VAL **in** READ(env(x),v) Δ **return** v) **in**
               def $v_2 = $ TAU Δ **return** 100 **in**
                TAU Δ **return** $v_1 < v_2$) **in**
      if v   **then** $S[x := x \oplus \underline{3}]$env
           else **exit** End-While   = (by the previous example)

trap [End-While → **skip**] **in**
  cycle **def** v = (**def** v_1 = (**choose** v: VAL **in** READ(env(x),v) Δ **return** v) **in**
                def v_2 = TAU Δ **return** 100 **in**
             TAU Δ **return** $v_1 < v_2$) **in**
        if v **then** **def** v = (**def** (v_1 = **choose** v: VAL **in** READ(env(x),v) Δ **return** v) **in**
                    def v_2 = TAU Δ **return** 3 **in**
                   TAU Δ **return** v_1+v_2) **in**
              WRITE(env(x),v) Δ **skip**
         else **exit** End-While
= b_2.

It will be seen that b_1 and b_2 are elements of $W_{Sig(STATE)}$, ie terms on the signature of the specification STATE hence we can see the denotational clauses as defining a translation from EL into the language $W_{Sig(STATE)}$; while considering b_1 and b_2 as abbreviations for their values in CALG, usually indicated b_1^{CALG} and b_2^{CALG}, we have the truly denotational definition (analogously for the direct semantics style).

3.3 *Semantics of the language*

Semantics as homomorphism. Having defined CALG, we have to show that the denotational clauses of section 3 define indeed a homomorphism. This should be clear since we have followed a traditional style; but, for sake of completeness and for indicating the pattern we have to follow when applying the method to other languages, we give some informal explanations of the technical result guaranteeing that denotational clauses are formally correct.

We consider the denotational clauses as defining a (homomorphic) translation (see examples of section 3.2) and denote by *P, E, D, S,* env, the corresponding functions and objects.

Under the assumptions that st, e, decs are statically correct w.r.t. the corresponding env, then we have to check that

   P[program st] ∈ $W_{Sig(STATE)}$|state,
ie it is term without variables of sort state on the signature of STATE; and

   S[st]env, D[decs]env, E[e]env ∈ $W_{Sig(STATE)}$|bh,
ie are terms without variables of sort bh on the signature of STATE.

Moreover all the above terms are provably defined in the specification STATE.

Interpreting terms in CALG gives the overall denotational semantics. The above assumption stresses the fact that our semantics can be seen as the composition of two homomorphisms: the first is a homomorphic translation into an algebra of terms on Sig(STATE), the second is the interpretation of these terms into CALG.

Operational semantics. There is an interesting alternative way of looking at our semantic specification, which we briefly outline.

First we can see, following the remark above the denotational clauses as a translation. Then we can associate to our specification of CONC_SYST its initial model I_{CONC_SYST}, which exists; it can be shown (see Prop. 1 in [ARW]) that in I_{CONC_SYST} s $\xrightarrow{f}$ s' holds iff it is provable in CONC_SYST. That means that I_{CONC_SYST} defines the operational semantics for CONC_SYST.

Finally it can be shown that, under some general condition, the congruence on $W_{Sig(STATE)}$ defined by CALG does correspond to the closure by congruence of an equivalence on programs defined from the transitions, a` la Milner (see, for the case of strong equivalence, [ARW]).

REFERENCES

(LNCS x stands for Lecture Notes in Computer Science n. x , Springer Verlag;
AdaFD n stands for: Deliverable n of the CEC MAP project: The Draft Formal Definition of ANSI/MIL-STD 1815A Ada.

[A] E.Astesiano, *Combining an operational with an algebraic approach to the specification of concurrency,* to appear in Proc. Workshop on Combining Methods (Nyborg, Denmark,1984), also in Quaderni Cnet n.127, ETS Pisa, December 1984.

[AGMRZ] E.Astesiano, A.Giovini, F.Mazzanti, G.Reggio, E.Zucca, *The Ada challenge for new formal semantic techniques,* in Proc. of the 1986 Ada International Conference, Edinburgh, Cambridge University Press, UK, 1986.

[AMRW] E.Astesiano, G.F.Mascari, G.Reggio, M.Wirsing, *On the parameterized algebraic specification of concurrent systems,* Proc. CAAP '85 - TAPSOFT Conference, LNCS 185,1985.

[AMRZ] E.Astesiano, F.Mazzanti, G.Reggio, E.Zucca, *Applying the SMoLCS specification methodology to the CNET architecture,* CNET - Distribute Systems on Local Network, vol 2, pp. 255-267, ETS Pisa,1985.

[AMRZ1] E.Astesiano, F.Mazzanti, G.Reggio, E.Zucca, *Formal specification of a concurrent architecture in a real project,* Proc. of ACM-ICS'85, North Holland, 1985.

[AR] E.Astesiano, G.Reggio, *A syntax-directed approach to the semantics of concurrent languages.* Proc.'86 IFIP World Congress, (H.J.Kugler ed.),p. 571-576, North Holland, 1986.

[AR1] E.Astesiano,G. Reggio, *Comparing direct and continuation styles for concurrent languages,* to appear in Proc. STACS 87' LNCS..., 1987.

[AR2] E.Astesiano, G.Reggio, *The SMoLCS approach to the specification of concurrent systems,* CNET - Distributed Systems on Local Network, vol 2, pp. 237-254, ETS Pisa,1985.

[AR3] E.Astesiano, G.Reggio, SMoLCS driven concurrent calculi, to appear in Proc. TAP SOFT'87, LNCS..., 1987.

[ARW] E.Astesiano, G.Reggio, M.Wirsing, *Relational specifications and observational semantics.* in Proc. MFCS'86 (Bratislava), LNCS 233, 1986.

[ARW1] E.Astesiano, G.Reggio, M.Wirsing, *A modular parameterized algebraic approach to the specification of concurrent systems,* technical report, 1986.

[ARW2] E.Astesiano, G.Reggio, M.Wirsing, *On the algebraic specification of function spaces,* in preparation.

[AW] E.Astesiano, M.Wirsing, *An introduction to ASL*, to appear in Proc. of IFIP TC2 Working Conference on Program Specification and Transformation (Bad Tolz, FRG) North Holland, April 1986.

[ASP] E.Astesiano, J. Storbank Pedersen, *An introduction to the draft formal definition of Ada*, in Proc. 3rd Workshop on ADA Verification, Triangle Park, USA, 1986.

[B] D. Bjørner, *The role and the scope of the formal definition of Ada*, AdaFD 4, 1985.

[BJ] D.Bjørner, C.B.Jones, *The Vienna Development Method:The Meta-Language*, LNCS 61, 1978.

[BO] D.Bjørner, O.Oest, *Towards a formal description of Ada*, LNCS 98, 1980.

[BW1] M.Broy, M.Wirsing, *On the algebraic specification of finitary infinite communicating sequential processes*, in Proc. IFIP TC2 Working Conference on "Formal Description of Programming Concepts II", Garmish, 1982.

[BW2] M.Broy, M.Wirsing, *Partial abstract types*, Acta Informatica 18, 47-64, 1982.

[BW3] M.Broy, M.Wirsing, *Algebraic definition of a functional programming language and its semantic models*, RAIRO Informatique Theorique 17,2, pp.137-161, 1983.

[CIP] CIP Language Group, *The Munich Project CIP*, LNCS n. 183, 1985.

[CRAI-DDC] E.Astesiano, C.Bendix Nielsen, N.Botta, A.Fantechi, A.Giovini, P.Inverardi, E. Karlsen, F.Mazzanti, J. Storbank Pedersen, G.Reggio, E.Zucca, AdaFD 7, 1986.

[H] H.Hussmann, *Rapid prototyping for Algebraic specifications RAP system user's manual*, MIP 8502, Universitat Passau, 1985.

[M1] R.Milner, *A calculus of communicating systems*, LNCS 92, 1980.

[M2] R.Milner, *Calculi for synchrony and asynchrony*, TCS 25, 267-310, 1983.

[Mo] F.Morando, *An interpreter for concurrent systems SMoLCS specifications*, Thesis (in italian) University of Genova, Italy, 1986.

[P] G.Plotkin, *A structural approach to operational semantics*, Lecture notes, Aarhus University, 1981.

[S] J.E.Stoy, *Denotational semantics: the Scott-Strachey approach to programming language theory*, The MIT Press, London, 1977.

[SW] D.T. Sannella, M. Wirsing, *A kernel language for algebraic specification and implementation*, In Proc. Int. Conf. on Foundations of Computation Theory, Borgholm, Sweden, LNCS 158, 1983.

[W] M.Wirsing, *Structured algebraic specifications: a kernel language*, habilitation thesis, Technische University Munchen, 1983.

PROJECT GRAPHS and META-PROGRAMS

Towards a Theory of Software Development

Dines Bjørner

Department of Computer Science
Technical University of Denmark

Dansk Datamatik Center
Lundtoftevej 1C

DK-2800 Lyngby
Denmark

22. August 1986

Abstract

Project Graphs are acyclic directed graphs and
define plans for the development of software.

Project Graphs are discussed from several Project Management and Software Engineering points
of view, from the point of view of Programming
Methodology, and also of Theoretical Computer
Science -- ie. from more informal, pragmatic
viewpoints, via formal methodological viewpoints, to a strictly theoretical viewpoint.

Finally the architecture of a Software Development System is outlined. Within a single, unifying frame it embodies the management, engineering and science of software development.

The work reported in this paper is partially
funded by the CEC ESPRIT RAISE project, partially by the Danish Technology Board, and partially by the Danish Research Foundation for the
Natural Sciences (SNF) grant no. 5.17.5.1.13.
(FTU)

0. BACKGROUND

0.1 Some Definitions

We start with some definitions:

By 'computer science' we understand the study of programs (viz.: complexity, computability, proof systems, automata theory, context free (etc.) languages, denotational and algebraic semantics), and the study of the mathematical techniques needed in studying programs and programming.

By 'computing science' we understand the study of programming (viz.: programming methodology, specification & design techniques, functional-, logic-, algebraic-, object-, and parallel programming).

By 'software engineering' we understand the practice of programs and programming, and the technologies of the various application fields (viz.: compiling, database, operating systems, computer networking techniques, etc.).

By a 'programmer' we understand a person who, in constructing software, follows the techniques and methods of computing science.

By a 'software engineer' we understand a person who supports programming and programmers in the context of changing technologies.

...

We elaborate upon the above definitions.

In computer (or program) science, programs are considered "passively": their properties are studied, but, to a first degree of approximation, not the process of constructing programs. It is studied what programs specify, not how they are constructed. This latter is in the domain of computing science.

The definitions may amount to hair-splitting in that the same person may work both in computer and computing science, or be both a programmer and a software engineer.

0.2 Programs and Programming as Formal Objects

Programmers are semanticists, i.e. they treat programs and programming as formal objects, and they form a bridge between theory and software.

Software engineers deal with syntax and pragmatics, and they form a bridge between programming and technology.

As we shall elaborate in this paper, programmers are theory builders -- in that programs denote theories, and software engineers are tool builders.

The major point of this paper is that of a particular proposal for a theory of software developments. It is based on the principle of considering programs and programming as formal objects.

Derivative points are those of refining our understanding of the

distinction between computer ·and computing science, of programmers and software engineering.

Engineering must, we believe, be based on science. And the tools of engineering most embody this science. We close this paper with the outline of a "machine" for the interactive execution of so-called project graphs, ie. for the extensive, semantics based support of software development.

0.3 Project Graphs and Meta-Programs

The present authors' notion of project graph expounded in this paper dates back to 1976. In [Bj|rner 77] the example of section 1. was first used, but only shown in its diagrammatic form in [Bj|rner 80]. In [Bj|rner 82] it was discussed at some length. The current presentation is more comprehensive.

The notion of project graphs and meta-programs is, however now in wider use. The author acknowledges inspiration drawn from [Scott 83] and [Sintzoff 84] Finally it must be noted that the various meta-program notions are not equivalent.

1. A WORKED EXAMPLE

We shall first motivate the reader by presenting, in this section, a worked example. Following sections (2-3-4) will then analyze the notion of project graphs more conceptually and will be presented in three parts.

1.1 The Compiler Project Graph

In this subsection we present a specific (applications-oriented) project graph. Without making a serious attempt we briefly motivate and explain the actions designated by the nodes of this project graph. This should, bearing in mind that we are dealing with a specific example, give the reader a feeling for the general kind of activities and relations between activities that can be designated by the nodes, respectively edges of project graphs.

The development of a compiler for a programming language like *Ada*, *CHILL, Modula-2, Pascal-Plus* or even as simple as *occam*, to a first approximation, consists of three parts: definition, design and implementation (or coding). But: within each part there are clearly distinguishable subparts. The following is a project graph specifying one kind of development of compilers for the class of languages including the above-mentioned and characterized by being ALGOL-like, ie. with statically decidable type (etc.) properties, but with concurrency.

l-**Fig.1** *The Compiler Development Project Graph* ------------------l

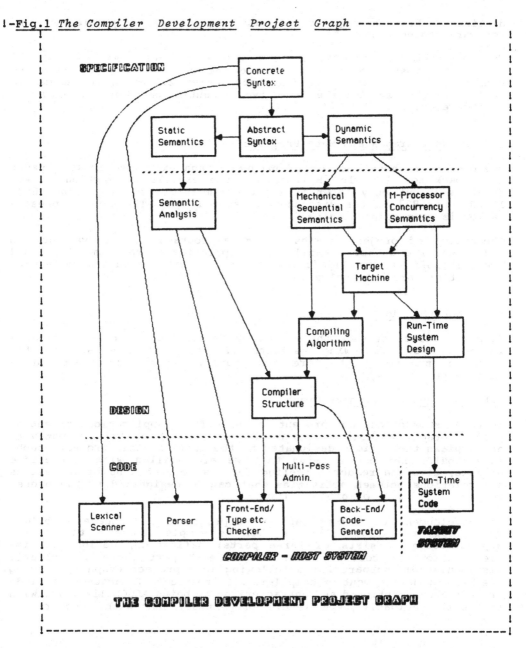

We now explain each labelled node and unlabelled sets of edges leading
into nodes. Their explanation is given in some topological order of
the acyclic graph.

1.1.1 Syntax

First we define the syntactic domains of programs (to be input to the compiler).

In the case of all of the above-mentioned languages the first node (with no input edge) would be the concrete syntax node. This is so since these languages would already have been given a BNF grammar before the developments, to which the present graph applies, takes, or took place.

From the concrete syntax we abstract a set of domain equations which focus at the most abstract level on which to look at input programs syntactically. We motivate this abstration by the ease with which we can subsequently define the semantics.

The edge leading from the concrete syntax to the abstract syntax node stands for the process of abstracting from the concrete to the abstract. The edge can be understood as an abstraction function from the concrete to the abstract.

1.1.2 Static Semantics

By static semantics we mean that meaning of programs which only takes their static properties into account, ie. properties which can be checked to hold without executing the programs (by providing them with input data and by running them). To the static properties we list: use vs. definition (checking that all identifiers which are used have been defined), type checking (checking that operands are of a type expected by operations upon them), uniqueness of distinctly defined identifiers, matching lengths, types, etc. of parameter/argument lists of eg. procedures (definitions vs. calls), overload resolution (annotating the exact type of operators applicable to distinct operand types), etc.

We motivate abstract definitions of the static semantics to ensure that the developer has a precise, consistent, complete and implementation-independent base from which to design the compiler front end. We suggest abstraction in order to get as short a definition as possible, and one that focuses on what the problem of static checking is, not how it is to be done.

The edge leading from the abstract syntax node to the static semantics node stands for 'inclusion', ie. the abstract syntax is part of a consistent and complete static semantics.

1.1.3 Dynamic Semantics

By dynamic semantics we mean the meaning of programs -- eg. as functions (over states) which compute results given input data. The dynamic semantics of a language focus on abstract run-time characteristics of programs.

We motivate abstract definitions of the dynamic semantics for reasons similar to the ones given at the end of section 1.1.2., but now for the development of the (host system) compiler back-end and the (target system) run-time support.

The edge leading from the abstract syntax node to the dynamic semantics node has the same meaning as that for static semantics, namely 'inclusion'.

1.1.4 Formal Language Definition

Thus: the two edges and the three nodes between which the inclusion relation holds together represents a project graph for the formal definition of a class of languages.

1.1.5 Semantic Analysis

The static semantics definition is normally too abstract to serve directly as the basis for coding -- and, if properly abstracted, usually cannot be directly executed. So a step (or more) of design is required. By a semantic analysis we understand the same as by static semantics. The change in words is meant, however to indicate that whereas the 'static semantics' (document) is intended to be formulated as abstractly as possible, or as is reasonable, the 'semantic analysis' (document) is to represent a reasonable design step of development towards an implementation. Exactly how concrete the design, at this level, should or cannot be fully edicted. It may be that several, sequentially stringed, increasingly concrete steps of design is needed. If so the 'semantic analysis' node should be replaced by such a strand of nodes. It should be emphasized, however, that we desire the (last such) semantic analysis node to result in a document which emphasizes <u>what</u> should be checked statically, and only abstractly, <u>how</u>. Thus the abstract 'static semantics' is here turned into an abstract algorithm for semantic analysis.

The edge from 'static semantics' to 'semantic analysis' stands for injection or abstraction: the former going from the abstract to the concrete, the latter from the concrete to the abstract.

We motivate abstract, rather than concrete algorithms for the reason that we desire to conquer complexity by carefully concentrating on a strategy for implementing the compiler front end type checker etc.).

As we shall see later we do provide for a step of concrete design.

1.1.6 Single versus Multi-Pass Compilers

It is in the design steps that we increasingly commit ourselves to some form of single phase or multi-pass compiler. Let us consider two extremes. If the host system has an indefinite addressing and foreground storage space, then (perhaps) a single phase compiler is acceptable. In this case we envisage that the internal representation of programs is such as to allow all of the program to be represented and present in the foreground store together with all of the compiler code which checks and translates the programs. If, on the other hand, only a limited, say typically a 128KB addressing space is at hand in the host system, if, furthermore, the compiler is required to compile indefinitely large programs (or separately compilable modules), and/or if the compiler code is expected to fill more space than is feasible, then a multi-pass compiler seems required.

By a multi-pass, eg. an N-pass, compiler we understand a compiler which can be viewed as consisting of N (or N+1) subprograms where the N subprograms are executed in sequence (as controlled by the N+1'st subprogram, the multi-pass administrator). A pass is defined as the processing of a complete program (or a separately compilable module) through a linear traversal of the program (or module). The program (or module) text, may be represented in some, possibly linearized, parse tree form. A linear traversal is then one which reads the tree bottom-up or top-down, and either left-to-right, or right-to-left, and, in doing so it may then perform a pre-order, some in-order, or a post-order tree-traversal!

1.1.7 Mechanical Sequential Semantics

The dynamic semantics is usually too abstract to serve directly as the basis for interpretation, let alone coding the code generator. So we conclude that one or usually more steps of concretizing design is needed.

By a mechanical sequential semantics we understand a dynamic semantics of the sequential, possibly still non-deterministic, aspects of the language -- and in such a way as to exhibit, in the abstract, the run-time organisation of programs and data.

Here we show just one node -- with the remarks of section 1.1.5 also applying here. We are assuming that the execution of concurrency constructs will be done by calls (of the generated code under execution) to some form of a target system run-time-system. We are also assuming that it is indeed possible to split the dynamic semantics into two: one for sequential, and one for parallel (or concurrent, tasking, or process) actions!

The edge from 'dynamis semantics' to 'mechanical sequential semantics' stand for a relation of injection, or, in the reverse, an abstraction function. In both cases preceded by some projection which filters away ("most") notions of concurrency.

1.1.8 M-Processor Concurrency Semantics

Usually the 'dynamic semantics' of concurrency is free from any notion of processes, or of there being a specific parallel-process scheduling principle. In the case of the compiler, we discussed, in subsection 1.1.6, a notion of single-phase respectively multi-pass compiling. We can now investigate a dual notion of executing programs with P processes on a target mono-processor, on a fixed (M-processor) target multi-processor, or on an indefinite (Q-processor, $Q \gg P$) target multiprocessor. We can also analyze w.r.t. whether the target multi-processors are tightly coupled, by sharing storage, or loosely coupled (distributed), each with their own storage, or some combination of these two forms. Finally the target operating system may or may not restrict allocation of processes to processors. Whatever is the case: we will in any case be confronted with a need to settle, in the abstract, for some process to processor allocation, scheduling, sharing and intercommunication principles. And this is then what we mean by an 'dynamic semantics' of concurrent processes, their execution, sharing and interaction and communication, from a free notion of processes (etc.) onto a fixed notion of processes (etc.)

The edge from the 'dynamic semantics' to the 'M-processor Concurrency Semantics' then designates first a projection which filters away ("all") aspects of sequentiality, and then some injection function from the abstract to the concrete.

1.1.9 The Target Machine

Either the physical target machine is given for which code is to be generated (like a VAX/11, a SYSTEM 370, a CDC CYBER 180, or other), or it is decided to generated virtual code, ie. code for an interpreter which can be easily ported among a reasonable variety of target machines.

In the former case we still need to formally define the specific architecture. In the latter case we can derive this architecture from the 'mechanical sequential semantics'.

In the former case we omit the edge from the 'mechanical sequential semantics' to the 'target machine semantics'. In the latter case this edge again denotes some injection abstraction relation/function.

1.1.10 Compiling Algorithm

By a compiling algorithm we understand a specification of a function from syntactic constructs of the source language to (sequences of) syntactic constructs of the target language. From the two specifications: the mechanical sequential semantics (of the source langurge) and the target machine language semantics we can now derive a compiling algorithm.

The compiling algorithm specifies **what** code the compiler (back end) should generate, **not how**.

Should a notion of optimization of run-time code performance be required, then that would likely require -- in the compiling algorithm step -- some notion of data and control flow predicates whose evaluation would determine code choices.

1.1.11 Compiler Structure

By 'compiler structure' we mean the detailed decomposition of the compiler into possibly multiple passes, the format of their intermediate texts, the structure and operations upon the dictionary name table, the design of separate compilation facilities, and the allocation of the **what** functionalites of the 'semantic analysis' and the 'compiling algorithm' to indvidual passes.

From the semantic analysis and the compiling algorithm we can derive final aspects and the possibly multi-pass decomposition of the compiler. Typically concrete versions of either kind of documents can be expresed in some extended attribute grammar. From this form we can formally derive the exact number of logical passes required to compute inherited and synthesized attributes by strict linear text traversals. Some of these logical passes, in code form, may still be too large to fit in the desired addressing and storage space. Then a further decomposition into physical passes is required.

1.1.12 Run-Time System Design

By a 'run time system design' we understand a design of the target system code which controls the execution of compiled programs. The control includes such items as storage (run-time stack and heap) allocation and deallocation, scheduling of individual processes of the executing program: their synchronisation and communication, the interface between the executing program and the target system I/O, etc. If the compiled code is not target machine code, but that of a virtual ("P-code") machine to be emulated, then the 'run-time system design' also includes the design of such a virtual machine interpreter (ie. emulator).

From the 'target machine' and the 'M-processor concurrency semantics' we derive the 'run-time system design'.

The edges leading into this 'run time system design' box has the meaning of injecting the "sum" of the 'target machine' and the 'M-processor concurrency semantics' into a 'run time system design'.

1.1.13 Design

Sub-sections 1.1.4 - 1.1.12 inclusive described the design efforts of the compiler development. Perhaps 'The Target Machine', section 1.1.9 should more properly be put under 'Formal Definition' -- together with that of the source language!

1.1.14 Multi-Pass Administrator

By a 'multi-pass administrator' we understand host system code which controls (sequences, etc.) the execution of all passes of a compiler, and which communicates intermediate texts between these and backing store.

The multi-pass administrator can now be coded from the specification of the compiler structure.

1.1.15 Lexical Scanner

By a 'lexical scanner' we understand host system code which assembles sequences of text characters into atomic tokens. (Keywords, some operators, delimiters, and user defined identifiers are often composed from sequences of 2 or more characters, but their syntactic and hence semantic significance can be expressed atomically given a sufficiently large universe of markers and tokens.)

The lexical scanner can usually be automatically derived from the concrete (BNF) syntax of the language.

The edge from the concrete syntax box to the lexical scanner box thus means: automatic derivation!

1.1.16 Syntax Parser

By a 'parser' we understand host system code which from a sequence of atomic tokens (ie. lexical objects) constructs a derivation sequence of syntax production applications. These would, from a non-terminal root lead to the atomic, terminal sequence. Usually the derivation sequence is represented in the form of some tree-like structure, the parse-tree.

From a deterministic BNF grammar one can usually automatically generate an error-correcting parser.

The edge from the concrete syntax box to the syntax parser box thus means automatic derivation.

1.1.17 Front End

By the 'front end' we mean (one or more) target machine independent host system code (passes) which implement the 'semantic analysis'.

From the "what" of semantic analyses and the "how" of the 'compiler structure' we derive the (one or more) code (passes) of the 'front-end'.

Thus the pair of edges from respective boxes ('semantic analysis' and 'compiler structure') to the 'front end' mean: the transformation (injection) of part of these former into the latter.

1.1.18 Back End/Code Generator

By the 'back end' we mean (one or more) usually target machine dependent host system code (passes) which implements the 'compiling algorithm'.

From the "what" of the 'compiling algorithm' and the "how" of the 'compiler structure' we derive the (one or more) code (passes) of the 'back end'.

And as for the pair of edges leading into the 'back end' box we give an explanation similar to the one given at the end of subsection 1.1.17.

1.1.19 Run-time System Code

By 'run time system code' we mean the target system code which implements the 'run time system design'.

The code is derived from the 'Run-Time System Design'.

The edge between these design and code boxes has the meaning of injection (from design to code), or abstraction (from code to design).

1.1.20 Implementation

Sub-sections 1.1.14 - 1.1.19 inclusive deals with the coding, or implementation aspects of the compiler development.

1.1.21 Conclusion

We have explained the individual steps in the specification, design and coding stages of the development of a compiler for a programming language with a fixed type concept, with imperative and with concurrency features.

The explanation has partially motivated some of the steps. The purpose of the current paper is not to argue that these steps must exactly be the ones followed for the kind of compilers we are seeking (although we might certainly wish to argue that -- elsewhere), but to illustrate the potential complexity of components of the development of large, complex software systems in general. The purpose has not been to advocate abstract specification, design, and concrete coding, but to illustiate that once you decide on this (recommendable) approach, then it is not just a matter of cascade or waterfall diagram simplification: specification - design - coding, but involves many more steps, not necessarily all stringed like beads on a string, but forming instead some acyclic graph!

1.2 A Review -- and some Comments and Management Remarks

We now review some aspects of the development of compilers. We do so in order to motivate sections 2., 3., and 4.

We also here bring som guidelines for management not found in sections 2., 3., and 4.

1.2.1 Resource Estimates

Given the project graph we can now attach estimates to relatively small, individually meaningful parts. Thus we are argueing that once the project graph has been established, then one can attach more realistic resource estimates: time and number of people needed.

These claims have been justified by the actual development, in the Dansk Datamatik Center (DDC), of both CHILL and Ada compilers.

1.2.2 Staff Mobility and Qualifications

Reviewing the project graph and based on actual experience we remark the following: the developers who formally define for example the static semantics need not also be the ones formally deriving the semantic analysis. In fact, we in cases advocate rotating developers so that for example developers who defined the 'dynamic semantics' formal ly derive the 'semantic analysis', etc. This has been tried in practice and been shown feasible and desirable. Thus, with the project graph, detailed at a sufficient level, in front of us we can better plan moving developers around.

We also remark that once for example the semantic analysis and the mechanical semantics has been constructed it is indeed feasible to use developers who need not know the source language in question. They need, ideally speaking, only be able to read and "understand" these documents -- which could be said to equal understanding the source language, but in fact is not equal to that! Thus the developers constructing the 'compiler algorithm' need, in principle, not know neither the source nor the target language!

The above: mobility and language aquisition are idealized situations. They are here phrased somewhat in the extreme, so as to make the point: with a well defined decomposition of the development task it becomes easier for management to manage!

1.2.3 Make Haste Slowly

In the case of the DDC Ada compiler the following very rough "statistics" can be given:

(1)   First one year, of a four year compiler development, was spent on formally defining Ada. In that year, ideally speaking, nobody considered the compiler. At the end of the first of four years nothing was, in principle, known about the compiler!

(2)   Then one year was spent on the "what" of the compiler and run-time system. After two years, again ideally speaking, nothing was known about the structure of the compiler!

(3)   Then half a year was spent only on the compiler structure: on
      the "how". After two and a half year no code had yet been written
      for the compiler!

(4)   Finally, in the last year and a half the compiler was coded and
      validated!

We believe the morale is: make haste slowly! We leave it to our rea-
ders to compare this sequence, and the fact that only a total of approximately 44 man years was spent on the whole affair! This is believed to be less than half of what leading competitors and, in promi-
nent cases, less than a quarter of what others spent on developing
their Ada compilers.

2. PROJECT GRAPHS AND SOFTWARE ENGINEERING

An example has been given of a project graph. In the next five sections we shall review this notion of project graph from various angles,
four analytical, one synthetical. In this section we review the
notion of project graphs from the viewpoint of software engineering.

2.1 Activities and Documents

Nodes

Nodes prescribe activities leading to specification, design, or code
documents. So, in a sense, nodes designate both activities, and documents.

The activities leading to documents are governed by a method: a set
of guidelines for carrying out the activities according to certain
techniques and using certain tools.

'Document' is a word covering both specification, design and code.
All results of software engineering and programming are (usually
electronically) recorded in the form of documents. The techniques and
tools used in creating documents give rules for when an activity is
finished, ie. when a document is consistent and complete.

If we choose to let a node designate a document which in itself is to
be a consistent and complete description of a specification, a design,
or a code, for some well-delineated part, or the whole, of a software
system, then the activities of a node usually consists of several
clearly separable sub-activities.

Let us take an example. In a one node denotational or mechanical semantics of some system (component), there are usually the creation
of the subparts mentioned in the following figure:

!—**Fig.2** *Sub-Node of a Single Node / Specification or Design* ——————!

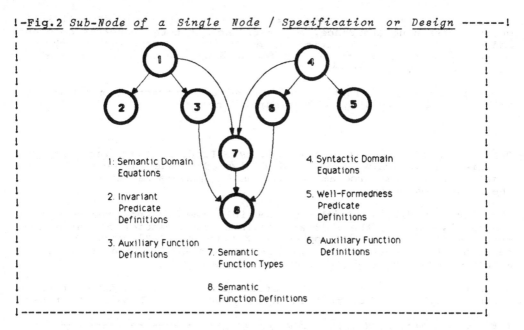

1: Semantic Domain Equations

2. Invariant Predicate Definitions

3. Auxiliary Function Definitions

4. Syntactic Domain Equations

5. Well-Formedness Predicate Definitions

6. Auxiliary Function Definitions

7. Semantic Function Types

8. Semantic Function Definitions

We choose not to decompose the project graph into further subgraphs like the above for several reasons. Firstly the form of the subgraphs is a property, not of the applications to be developed, but of the technique used. Secondly the graphs, as a picture to be read, would be too complicated. The first reason is purely a semantic one, and is sufficient. The second reason is purely a pragmatic one, but is necessary!

So we conclude that for each node there is a method with its techniques, for carrying out the activities of the node. And the possible decomposition of the node into a subgraph is a property of the method, and is henceforth to be treated as part of the software engineering "meaning" of the node!

2.2 Correctness and Validation

-- The Contractual Model of Development

A pair of nodes with a connecting edge also symbolizes a contract. A contract between those who developed the abstract node and the concrete node documents. This contract involves both verification of correctness and validation of additional, typically non-functional requirements -- for example such embodied in man-machine interfaces, time-dependency properties, space/time consumption (ie. efficiency), robustness, reliability, fault-tolerance, security, etc.

By 'verification' we mean some sort of rigorous argument of correctness: that the concrete definition (under some equivalence criterion) expresses the same as the abstract definition. Verification may entail formal proofs.

By 'validation' we in general mean verification plus something more. Here we mean just "that something more". "That something more" may

130

include 'testing', "walk-throughs", etc. By this form of validation
we mean obtaining assurance, or confidence that the concrete document
indeed meets its non-functional requirements. 'Testing' is then a
systematic and organized search for a counter-example to the claim
that a document is correct.

The above three paragraphs have implied a 'contractual model of development' which we diagram in figure 3, below.

o o o

Once the High Level Design produced by a Producer (Supplier) has been
verified and validated then it is delivered to the customer (who con-
sumes it). The same programmers/software engineers may be customers of
their own productions!

The figure 3 diagram, with its possibly re-iterated productions of
drafts, is part of the software engineering semantics of three con-
secutive nodes; that is: boxes *1-2-3-4* (or *4-5-6-7*) with their in-
ter-connecting edges is part of the software engineering semantics of
a pair of adjacent nodes of a project graph.

We do not show the intermediate "draft" and "verify/validate" nodes in
Project Graphs as they pertain equally well to all pairs of nodes.

l-<u>**Fig.**3</u> *Contractual Model of Development* ------------------------l

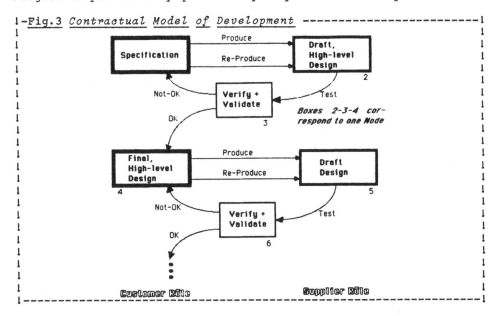

More precisely boxes *2-3* (or *5-6*) with the edges leading to, from
and between them is part of the software engineering semantics of any
edge.

2.3 Animation and Prototyping

By 'animation' we mean the "direct" execution, symbolic or otherwise
of a specification or design. By "direct" execution we mean that some

automatic device (a program + machine) obeys the specification or design. By symbolic execution we mean an execution where some or all user-defined identifiers of the specification or design are not bound (or evaluated) to a value (but left free).

By a 'prototype' we mean a partly or wholly manual derivation of a high level executable design from a specification (resp. a design from a high level design, etc.).

More precisely: we think of animation as something done by and on behalf of the software engineers and programmers who produced the animated document. And we think of animation being a process only applied to a document, free from any environment (e.g. the one into which the product eventually is going to be embedded).

Correspondingly we think of prototyping as something done by the software engineers and programmers who produced, or who represents the producers of, the prototyped document, but on behalf of some customer. And we think of prototyping as being a process which properly embeds the prototype in an environment identical to or typical (a prototype) of the one in which the product will be embedded.

Animation and exercising prototypes is typically done in order to investigate non-functional properties of the specified or designed system.

(Testing can potentially be systematically performed on prototypes, or through animation. Testing is one way of investigating properties.)

The 6 topmost nodes (3 in "column" 1, 1 in "column" 2, 2 in "column" 3) and the edges between them of the following diagram therefore gives software engineering meaning to project graph nodes:

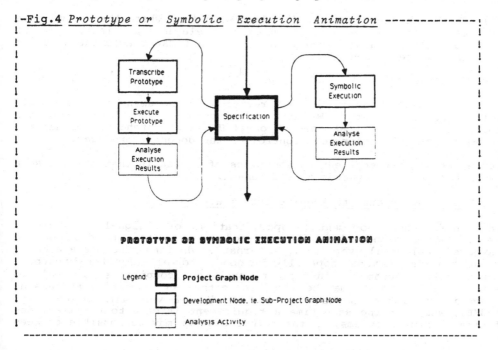

I-**Fig.4** _Prototype or Symbolic Execution Animation_

PROTOTYPE OR SYMBOLIC EXECUTION ANIMATION

Legend:　☐ **Project Graph Node**

☐ Development Node, ie. Sub-Project Graph Node

☐ Analysis Activity

2.4 Versions and Journaling

-- Versions

For every node a document is created. In fact, we may find that for every node, several versions of "similar" documents will be created.

By a 'version' of a document we understand just a document which otherwise, as all documents must, satisfies its requirements, correctness and validity. So the word 'version' covers a pragmatic concept, where the term 'document' covers a semantic concept. An example may illustrate our point: in developing our compiler we may wish to re-target (ie. develop) the "same" compiler for different target machines. For each such we develop a separate 'compiling algorithm'. So the node of that name may stand for several versions of (basically) "the same (!) document". This would entail both several versions of the 'target machine' document, and of the 'back end/code generation' document!

So the engineering semantics of any node is that it stands for a set of versions of "the same" document.

-- Journaling

When creating documents these undergo a number of states:

(I)   There are the stages where the document is known to be incomplete: it is being input, or previous input is being revised.

(U)   There are the stages where it is hoped to be consistent and complete but where tests, verifications, validations, etc., show otherwise -- as a result of which the document is being changed (updated).

(R)   There are the stages, coincident with (U), where the document is being improved: either to become more elegant, succinct, etc., or to reflect changing requirements -- in these stages the document is also being updated.

etc.

It is our attitude that no input revisions or document updates should unintentionally be lost. We are of the opinion that "old" documents could be as "correct" as "newer" ones! Hence we intend to journal all changes, possibly under semi-control of the document producer.

Hence yet another engineering semantics of a node is that it is also a repository for all updates to a document.

2.5 Design Decisions and Requirements Tracking

In transforming a document (a specification, or a high-level design), ie. in going along edges ("executing" edges), certain designs are favoured (selected) over other, alternative designs. Choices are made. These choices are made, hopefully for good, and well-identified reasons. Let us take an example: in deciding upon the 'compiler structure' two extreme design points may be possible: either an indefinite addressing space is available to the compiler, or just a very limited one (eg. 128KB), while at the same time a requirement is made to compile indefinitely large programs. In the former case it may be possible to keep both the (entire) compiler and very large (separably compilable sub-) programs in storage - - implying the possibility of a one phase compiler (no overlaying, co-routining etc. of compiler parts, and no intermediate spilling of texts (program parts, possibly transformed) onto backing store). In the latter case a multipass compiler seem required.

So the design decided upon hinges upon requirements of the host system.

For a set of given explicit requirements there will usually still be design decisions whose selection hinges on implicit, or expected "requirements".

An example is: there are many ways of implementing, at run-time, the calling of and passing of parameters to procedures. A design choice is made among these dependent on the expected use of procedures and of the patterns of references made in procedure bodies to local, (lexicographically) intermediate and global (outermost) blocks!

So the engineering semantics of an edge of a project graph, ie. of going from a node to a node, entails that of properly recording design decisions and of tracking these back to requirements definitions!

Design decisions are made at various stages of the development. A project graph should be annotated, usually from the time of its conception, before actual development starts, as to where, along which edges, which requirements will imply taking decisions. We do not show these "points" -- we could, by for example changing edges:

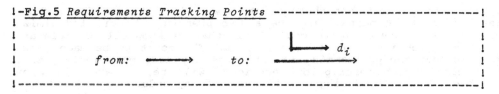

|-Fig.5 *Requirements Tracking Points* --------------------------------|

$$from: \longrightarrow \qquad to: \quad \Big\llcorner\!\!\longrightarrow d_i$$

where d_i then refers to some requirements and decisions.

Since we expect almost every edge to constitute such decision points, we take it as a part of the engineering semantics that edges always (normally) has such decision points.

2.6 Configurations

A project graph is a schema: as a software engineering device it describes a whole class of softwares. The example graph of section 1. for example can be used in the same syntactic form for the development of such widely different languages as Pascal Plus, CHILL, Ada and occam. "Applied" to one particular language, say Ada, a so instantiated schema may still describe a variety of ("similar") Ada compilers, for example all with their distinct 'compiling algorithm', 'compiler structure', and 'back end code generator'. Thus such an instantiated project graph, through permittable combinations of versions of node documents, describe several distinct configurations.

Thus a software engineering semantics of project graphs is that they also permit, through suitable annotations of node (and some kind of "overlay" graphs), the definition of a class of software configurations, and hence of their control.

3. PROJECT GRAPHS and PROGRAMMING METHODOLOGY

The software engineering properties of project graphs, ie. of software development, are typically syntactic and pragmatic. The programming methodology aspects of project graphs will be semantic.

3.1 Description Techniques

By a 'description' we mean either a formal definition, a specification, or a design (document). By 'descriptional techniques' we mean:

(1) The style of description -- be it denotational, algebraic, axiomatic, operational or mechanical, or some composition thereof.

(2) The structuring and modularity principles used -- expressing specialisation, generalisation, enrichment, parameterization, etc. of parts of descriptions, or applying various other subpart-composition operators.

(3) The technicalities of describing technical issues -- eg. such as using either continuation, exit, or direct semantics in modelling GOTOs of programming languages, or atomic (1-level) or structured (composite) locations when modelling array or record variables of programming languages, or using imperative or applicative (functional) style description when modelling (sequencing in) imperative systems, or using (or not using) polymorphic functions when expressing manipulations of type-expressions, locations and values, etc.

It should be obvious that there exists a wealth of description techniques, and that it perhaps may be a bit premature to try taxonimize these now -- since doing so could imply that we know all possible and applicable techniques. Be it far from so. The point to be made, however, is that the programming semantics of any node must include that it be clear which description techniques will be, or are being, or have been used in producing the document of any node.

3.2 Design Derivation Techniques

By a 'derivation' we mean a relation between two documents, of nodes connected by an edge. A design derivation is hence a part of the programming methodology semantics of an edge. A derivation "goes from a more abstract to a more concrete description".

A number of derivation schemes are possible:

(1) Object transformation, refinement, and enrichment

(2) Operation transformation, refinement, and enrichment

(3) Theory shift -- for example from an applicative to an imperative description, or from a description to be understood in a full Scott theory, to one understood ("only") in a naive Blikle theory, etc.

(4) Institution shift, a shift where two different logical systems lie at the foundation of respective nodes -- for example one (e.g two-valued) logic based on total functions, to one (e.g. three valued) logic based on partial functions.

So the programming semantic of an edge must include applicable derivations.

3.3 (Other) Inter-Node Relation Technques

Edges also designate other relations between documents than the above derivations. As such edges prescribe activities which lead to documents.

Edges go from nodes to nodes. The "from" nodes we (relatively speaking) call the abstract node (definition, etc.), the "to" node we call the concrete node (etc.)

Typical such relations are: injection, abstraction, adequacy, or inclusion. By 'injection' we mean an activity which embeds an abstract definition in one of several possible concrete representations. Injection is a relation: several "mappings" from (one) abstraction to concretisation are possible. By 'abstraction' we mean an operation (a function) which retrieves an abstract definition from a concrete one. By 'adequacy' we mean an assertion which expresses that to each abstraction, i.e. to each object (and operation) in the abstract world there corresponds at least one (non-trivial) object (and operation) in the concrete world. By 'inclusion' we mean that the target node document in order to be complete includes that of the source node.

So the programming semantics of an edge must include the techniques for expresing these relations.

3.4 Compositions

By 'composition' we mean the act of describing a specification or a design using two or more description lanquages.

In describing a system (a lanquage, etc.) we may find that one description lanquage is insufficient -- that no single such language has the appropriate features to handle all aspects of the system: both deterministic, non-determinstic, concurrent and time-dependent. We may therefore have to resort to using two or more lanquages.

We may for example use trace assertions to describe externally observable behaviours of a set of concurrent processes, CSP to describe their internal properties, and temporal logic to describe fairness properties.

The programming semantics of a node therefore must include that of the composition techniques used, if at all.

3.5 Proof and Verfication Techniques

In developing a description there may arise the need for checking that it satisfies certain properties.

Hence the programming semantics of a node must include a proof system for the description language, and eventually the properties to be proved and their proofs.

In tranforming one (abstract) description into another (concrete) description there usually arises an obligation to prove the correctness of the derivation. Many derivations will be a-priori correct, but there will always be cases where a proof is needed.

Hence the programming semantics of an edge must include a proof system for the derivation system, the description languages used on either side of the edge, and eventually the theorems to be proved, (sketches of) strategies and tactics to be used in the proofs, and the proofs themselves.

(In the paragraphs 1 and 3 before the present one the word 'eventually' separates those part of the node and edge semantics which should be present before actual development start from those which will transpire as a result of the development.)

We do not expect to be able to actually prove complete developments correct! But we expect theorems, and lemmas, and sketches of strategies and tactics to be provided by the development of interesting, but difficult extracts of the descriptions and the derivations. These extracts may be termed "ideal", but chosen carefully they should help increase considerably confidence in the correctness of the entire software.

4 PROJECT GRAPHS AND THEORETICAL COMPUTER SCIENCE

We shall now attempt to establish the exact, mathematical meaning of a project graph, its nodes and edges.

4.1 Nodes Denote Theories

From one kind of software engineering viewpoint nodes designate documents. From a programming viewpoint these documents are presentations of formal systems, be they abstract or concrete. From a computer science viewpoint a formal systems defines a theory. A theory, in mathematical logic, is a set of axioms, deduction rules and all the (interesting) theorems the former define. A formal specification or design define a set of system (for example programming language) properties, ie. theorems.

Examples of theories and meta-theories relating to the example of section 1. will now be given.

The 'concrete syntax' (of fig.1) typically designates a BNF grammar, ie. a context free grammer, and thus denotes a theory in the meta theory of context free languages.

The 'abstract syntax' (of fig.1) usually designates a set of usually recursive domain equations, and denotes a theory in the Scott meta theory of reflexive domain (structured as lattices or neighbourhood systems, and admitting only continuous monotonic functions (between domains)).

The 'static semantics' (of fig.1) can be presented in a number of ways: as an SOS (structured operational semantic) specification, as a denotational semantics, as an extended attribute grammar, etc. In each case the designated specification identifies a theory and a (well-known) meta theory.

The 'dynamic semantics' (of fig.1) can likewise be presented in a number of ways: as an SOS, a SMoLCS, a D-SMoLCS, a Meta-IV/CSP etc. specification -- again identifying respective theories and (well-known) meta theories.

The 'semantic analysis' (of fig. 1) can conceivably be presented as a functional semantics which can be understood either in a Scott theory of retracts, or in a (de-sophisticated, naive) set-theoretic setting. We may in fact choose to use the very same notational system to cover several theories. In this case the node theoretically denote a pair of theories and mappings (or shifts) between them.

The 'mechanical sequential semantics' (of fig.1) may typically correspond to a theory generated by a SMoLCS theory, a Metric- or a Set-theoretic Denotational Semantics theory.

The 'target machine' (of fig.1) can be expressed in the same meta theories as mentioned in the two previous paragraphs and themselves denote theories.

The 'compiling algorithm' (of fig.1) can typically be expressed in a set-theoretic denotational semantics, with presentation being easily (automatically and trivially) translated into an attribute grammar, ie. a presentation in some other meta theory, namely that of extended attribute grammars.

The 'lexical scanner' corresponds to a theory in a finite state machine (or automata) meta theory, and the 'syntax parser' to a theory push down stack machine meta theory.

The code of the 'front end' or 'back end code generator' can be understood in many meta theories. If for example coded in Pascal for which a Hoare (-like) proof system is given (and used), then we can claim that these code nodes denote theories understood the meta theory of a Hoare proof system.

Much work still remains to be done to identify the proper theories and of ensuring that actual developments of nodes do indeed take place within such theories.

Institutions

An institution is roughly speaking a collection of theories within the same logical system. Examples of logical systems are first order logic (FOL) without equality, first order logic (FOL$_E$) with equality, Horn clause logic (without equality), etc.

Several of the theories interconnected by edges may belong to the same institution. But usually not all will.

So a project graph also defines a set of (not necessarily disjoint) subgraphs, each belonging to one institution, ie. a partition on institutions.

4.2 Edges May Denote Morphisms

From a programming methodology point of view an edge prescribes a derivation of one document into another. From a computer science viewpoint such a (syntactic) derivation denotes a morphism.

If the adjacent theories belong to the same institution the morphism is a theory-morphism, whereas if the nodes belong to different institution the morphism is an institution-morphism.

Where two or more edges are incident upon one node the morphisms have suitable projection and composition properties on the cartesian product of the originating theories.

Much work remains to be done to clearly identify exactly the nature of the various morphisms.

4.3 Graphs May Denote Categories

Research, by my student Mogens Nielsen, is going on studying the kind of categories denoted by project graphs.

4.4 Meta-Programs

So, by a project graph we in a computer science setting understand a syntactic object, and by a meta-program we understand the mathematical (catefory theoretic) object denoted by it.

5. PROJECT GRAPHS and MANAGEMENT

Reference is made to section 1.2 where a number of remarks are made concerning such project management issues as: resource estimation, staff allocation and staff qualification.

Based on these remarks and the contents of sections 2-3-4 we summarize their management implications.

Our base dogma are these no software development without first a project graph. All development facets must be described by the project graph.

We decompose the concept of project management into the following technical facets: direction, control and monitoring.

5.1 Project Management Directives

We see it as the task of project management to construct major parts of the project graph. If management does not know which project graph to construct then it is our advice to not undertake any development! So we expect management to do the "programming" which leads to a suitably annotated project graph (see section 6.2.1). If management is not itself capable of identifying necessary and sufficient steps of development, the techniques to be used in each step (node or edge), and their underlying theories, then one should not embark on the development! And likewise: management should itself be qualified to carry out the development steps. So the major management directive lies in it determining the project graph.

5.2 Project Management Control

After management itself has "programmed" the overal project graph it now controls its execution. That is: management decides when execution have finished, and it decides on which resources should be allocated ato such node and edge executions.

5.3 Project Management Monitoring

In order for management to decide on when node and edge development activities (executions) can start and are finished management must be able to monitor progress: that is to check completion status of sub--activities within node and edge executions.

5.4 Conclusion

These three aspects: project graph construction, execution control and document status interrogation (monitoring) are all provided for in the support system now to be unvieled.

6. A SOFTWARE DEVELOPMENT SYSTEM

Software development support systems should support methods of programming, engineering and managing software development. The APSE (Ada Programming Support Environment) basically only supports Ada programming, and not even then the methods that may underly such programming. APSEs and other support environments primarily, if not exclusively only supports the software engineering, ie. some pragmatic and most syntactic aspects of engineering.

Project graphs, as we have now defined them, are a model for a software development methodology, and this model embodies most relevant engineering, programming and management aspects. Project graphs are based on a theory, althouth they were first conceived in practice. Most, if not all support environments do not have an underlying theory.

We shall now outline a system for supporting the engineering, the programming and the management of software developments based on project graphs.

6.1 Overview

Very briefly speaking: the software development system, together with the engineer, programmer and manager, interactively executes project graphs, by executing nodes and edges, possibly concurrently, due to the simultaneous activity of several subgroups of people.

So there are engines within the system which executes nodes, and other engines which executes edges.

To execute a project graph such a graph must first have been input to the system. So there is an engine of the system which reads, updates and outputs project graphs, and which delivers such for execution by the afore-mentioned engines.

The above three engines constitute the programming and some of the engineering aspects of the system, and that group we call the programming and engineering processor.

To manage developments based on project graphs the manager must be able to summarize the state of project graph based development: the completion status of the graph itself and, when subject to execution, the status of (or state of programs within) nodes and edges. So there is a processor within the system which provides management with not just monitoring facilities (as only mentioned above), but also control facilities over developments. We call this the management processor.

The collection of the programming and engineering and the management processors form the project graph machine.

To insert new tools, and to accomodate for new theories and institutions a tool and theory machine is provided.

The terms machine, processor and engine terms are intended to give the reader and user the feeling of something "mechanic".

So we may summarize the above in for example the below diagram.

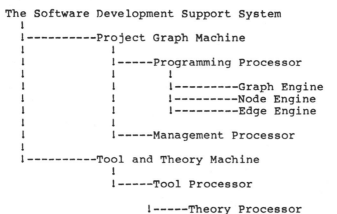

```
The Software Development Support System
 |
 |----------Project Graph Machine
 |              |
 |              |-----Programming Processor
 |              |        |
 |              |        |---------Graph Engine
 |              |        |---------Node Engine
 |              |        |---------Edge Engine
 |              |
 |              |-----Management Processor
 |
 |----------Tool and Theory Machine
                |
                |-----Tool Processor

                |-----Theory Processor
```

(This diagram differs slightly from the *Software Development System* figure at the end of the paper -- page 38.)

Within the node, and edge leaf engines, and within the tool and theory leaf processors, we shall decompose facilities into such which deal with syntactic and pragmatic (ie. engineering), respectively semantic (ie. programming) concerns.

6.2 The Graph Machine

This machine supports the development and execution of project graphs.

6.2.1 The Graph Engine

This engine supports the development of project graphs.

To develop a project graph first the graph must be drawn, and then its nodes and edges annotated.

To annotate a node is to attach information of the following kinds to the node:

-- Node Names

A (variable, free) node name is a purely pragmatic object. It should be assigned to a node before the graph is ever executed. It serves no other practical functions than to give a name to the activity and document of the node. It is adviced, again for purely pragmatic reasons, that no two node names of a graph are the same.

-- Description Style and Theory Names (Sects. 2.1, 4.1)

A constant name out of a fixed vocabulary of names of description styles must also be given to a node before any execution of that node can take place. Description style-names typically are: BNF grammar, Scott (or Blikle) Domain Equations, Meta-IV Standard (or Direct, or Continuation, or Exit) Denotational Semantics, OBJ/CLEAR, ASL, LARCH, SOS, SMoLCS, D-SMoLCS, CCS, Temporal Logic, (Extended) Attribute Grammars, Pascal, Ada, etc.

These description style names identifies both the notational system to be used, the specitic variant of description -- as in the case of direct/standard/continuation or exit denotational semantics --, and the underlying meta theory.

-- Animation or Prototyping (Sect. 2.3)

An indication can be given, either during development of the project graph, or during its execution, of the possible kinds of animation and/or prototyping experiments (or analysis) that a given node may/could/should give rise to.

-- Versions and Configurations (Sects. 2.4 & 2.6)

From before any development, to and including the final coding of an initial, or subsequent versions of the software, annotations may be made concerning versions and/or configurations of systems or subsystems.

Recall that each node may give rise to more than one ("similar") document (ie. version). For a given project graph (with all documents completed for all nodes) certain combinations of versions constitute a valid configuration.

A version annotation is simply information attached to nodes. A configuration annotation is an instance of a project graph with exactly one version document for each node.

-- Journaling (Sect. 2.4)

During the execution of a project graph errata in and updates to node and edge documents will be recorded, and past "text" will be journaled. The appropriate nodes and edges of the project graph should be automatically updated with respect to their journaling status to reflect this document history.

-- Requirements (Sect. 2.5)

Before execution of any of the non-initial, or, more narrowly, the design nodes, external, non-functional requirements must be established. Each such requirement will in turn be dealt with through the taking of certain design decisions. These will be taken at certain edges. Hence we propose to attach the totality of external non-functional requirements to edges of the project graph: zero, one or more per edges -- with no two edges being annotated with the same requirement.

Hence from a complete and consistent project graph one can extract an external, non-functional requirements document!

-- <u>Descriptional Techniques</u> (<u>Sect. 3.1</u>)

During project project graph development, or at the latest, before
node execution, annotations must be made as to which combination of
descriptional techniques will be (or are expected to be) used.

-- <u>Derivations</u> (<u>Sect. 3.2</u>)

During actual development, ie. during execution of nodes and edges the
programmer decides on certain derivations. The name and description of
the techniques used must be recorded, and will be attached to appropriate edges.

-- (<u>Other</u>) <u>Inter-Node Relations</u> (<u>Sect. 3.3</u>)

Similarly during development the recording of the use of other relations between documents are deemed essential -- and their name and a
description of them will be attached to appropriate edges.

-- <u>Description Compositions</u> (<u>Sect. 3.4</u>)

(This is a special case of description techniques, sect. 3.1.)
For each document various parts, or all of it, can be said to be
based on a view of the system that is either deterministic, non-deterministic, concurrent, or time-dependent. In planning development, ie. before actual execution of the project graph it should
be recorded, per node, which one, or a combination, of these views
prevail, or are to be emphasized by a document description!

-- <u>Proof</u> <u>and</u> <u>Verification</u> <u>Obligations</u> (<u>Sect. 3.5</u>)

When a derivation has been decided upon (as early as during project
graph development) (as late at during edge and ("next") node execution)
an obligation to prove the correctness may arise. The fact that such
an obligation is present and the way in which the obligation is (to
be) disposed off must be recorded. So should the (name of the) proof
system to be used, and an instructive description of the strategies
and tactics actually used in establishing the proof. These recordings
are to be attached to appropriate edges, and must also include an informal description of the sub-system(s) proved.

Finally concerning a node document one may wish to verify that it
satisfies certain properties. Again, as for above, the appropriate
nodes must record which properties ought be, or has been verified,
the applicable proof system, strategies and tactics.

Conclusion

To plan a project is to program. The programming of a software development consists of creating its project graph and in deciding, during
programming and during execution upon each of the 11 points made above.

Once a project has completed, ie. before its results ("the product")
are handed over to system installation, operation and "maintenance",
its complete documentation consists of (1) the completely annotated
project graph, (2) all the documents referred to by nodes and edges,
and (3) the development support system needed to "maintain" the product.

The root of all documents is the project graph. All other documents are leaves w.r.t. to this root. The development support system underlies these documents.

The Graph Engine does not merely record and maintain the graph and the 11 point annotations, it also performs a number of consistency and completeness checks and provides output (interface) to the remaining programming processor parts: the node and edge engines, and to the management processor. These will be outlined next. To perform the necessary consistency, completeness, and other checks, the graph engine needs input from the Theory Machine.

6.2.2 The Node Engine

The node engine provides services to the programmer and software engineer in developing the document(s) associated with a node. We group these services in the three categories of:

```
Services
    |
    |---Syntactic
    |---Pragmatic
    |---Semantic
```

The syntactic facilities put at the disposal of the software engineer have to do with the form of the documents, and the pragmatic with the use of the documents in the particular context of the given project. The semantic functions available to the programmer deal with the content of the documents.

6.2.2.1 The Syntactic Parts

We group the syntactic facilities in two: those which have to do with the form of input and output, and those which have to do with the form of journaling. The use of the syntactic facilities is basically independent of the specific project being pursued.

6.2.2.1.1 Editing and Formating

Editing: The node syntax engine provides for every notatonal system a variety of concrete input forms, forms that can be tailored to the individual engineer, ie. input forms that need not adhere to a standard concrete syntax for input, only to an abstract such. And the node syntax engine provides these input forms through the media of ordinary full screen editors, syntax directed editors, and more sophisticated structure editors. All of them interactive.

Formating: The node syntax engine correspondingly provides for a variety of "pretty printing" facilities: onto paper, and onto screen: both line/character oriented, as well as plotter printers and bit-mapped displays. And within each of these formating is provided on a customer-dependent, tailor-made way.

So common to both input and output is the adherence only to an abstract syntax, with facilities for tailoring concrete forms to the individual developers, respectively customers.

6.2.2.1.2 Journaling

Incremental input (updates and changes) to documents' are saved for eventual restoration -- either as a result of back-ups due to machine failures, or as a result of the engineer "regretting" later input in favour of earlier ones. This input journaling can be performed either automatically, or under engineering control, and can be either line/word/character oriented, or structure oriented.

6.2.2.2 The Pragmatic Parts

We group the pragmatic facilities in three:

1. those, at a **higher level**, which compose documents associated with nodes of the (entire) project graph into configurations;

2. those, at a **medium level**, which relate parts of documents to parts of other documents for the sake of tracing description decisions to earlier ones and to requirements;

3. and those, at a **lower level**, and on a per document basis, which annotate the documents for ease of understanding.

6.2.2.2.1 Versions and Configurations

To each node there basically corresponds one (the archetype or generic) document. This document may be "virtual" in the sense that it only "exists" in the form of all it versions, each version being an instatiation of this virtual, or generic/archetypal, document. (We refer to sections 2.4.1-2 for a motivation, and background.) So to each node in a project graph there may exist several (at least one) documents, and to a project graph there may then exist several instantiations of configurations, with a configuration amounting to a project graph with exactly one document (one version) per node.

The node pragma engine provides facilities for defining, creating, monitoring, and controlling versions, for relating them to configurations, and for defining, monitoring, controlling and generating configurations.

6.2.2.2.2 Recording Architecture and Design Decisions

In choosing specific levels of abstraction in specifications one is trying to emphasize certain characteristics of the evolving software architecture. Informal, natural language text annotations serve to focus on such specification decisions -- which in their formal appearance usually "hide" their pragmatics.

In expressing a design one is normally choosing one design over other likely designs, but, in the formal document, only one is recorded. Informal, natural language text annotations -- possibly supported by formal parts describing aspects of other design possibilities -- can then serve as a "repository" where the rationale for the specific design chosen is given, and where other contemplated design possibilities are "remembered" with reasons given for their rejection.

The node pragma engine provides facilities for annotating documents with respect to the choices they reflect, and with respect to the alternatives, which they do not reflect. The node pragma engine also provides facilities for tracing design choices and alternatives through a project graph for the purpose of checking that requirements have been fulfilled, or for the purpose of later evaluations which reconsider design changes in the light of changing requirements.

6.2.2.2.3 Annotating Descriptions

Formulae are just that: formal. Often they "express" their meaning in ways which hinder an easy understanding. We take as a dogma that basically every line of every document must be annotated in the form of natural, application-oriented, profession-language. The purpose of these annotations is, independent of the design decisions, to "read" the documents.

The node pragma engine provides facilities for annotating documents, and for extracting, from documents only these annotations for the purpose of creating informal introductions and reference manuals.

6.2.2.2.4 Requirements Tracking (*text omitted*)

6.2.2.3 The Semantic Parts

The purpose of the node semantic engine is to provide for a variety of abstract interpretations of formal documents. From a pragmatic point of view we group these partial computations in four areas, but stress that they all require basically the same underlying mechanisms for effecting these computations.

6.2.2.3.1 Static Checking of Documents

Usually the notational systems (specification and design languages) are "more or less strongly" typed, and otherwise impose a number of contextual restrictions which cannot be checked for even in a syntax directed input phase.

The node semantic engine provides for the static checking of such documents. To illustrate, these checks may include: operands are of the type expected by operations upon them, use of defined quantities adhere to their definitions, etc.

6.2.2.3.2 Dynamic Animation of Documents

To animate a document is to perform a computation covering some facet of what the document describes. Static animations were covered just above, in sect. 6.2.2.3.1. A dynamic animation corresponds to an attempt to actually compute the function (values) described, or some such similar thing! As such a dynamic animation corresponds to a concrete execution of its semantics. Now descriptions (specifications and designs) may be abstract to a degree which in general effectively precludes "actual" executions. In such cases one may still wish to perform symbolic computations, ie. executions which "rewrite" the text of the descriptions into other texts, texts in which some of the identifiers of the description document have been bound to values, others not.

The node semantic engine provides for a variety of such animation capabilities, all very much dependent not only on the level of abstraction language used, but also on the specific description given. Symbolic animations will be highly describer interactive.

146

(Section 2.3 motivates the animation capability.)

6.2.2.3.3 Prototyping of Documents

The node semantic engine provides functions for the interactive, developer controlled (monitored), and not necessarily automatic transcription of specification and (high level) design documents into executable documents.

(Examples of target languages could be languages like Standard ML, SETL and PROLOG.)

(Section 2.3 motivates the prototyping capability.)

6.2.2.3.4 Proving and Verifying Document Properties

The node semantic engine provides sub-systems for the interactive developer controlled verification of developer or customer stated properties of documents, ie. of the abstract software described by the documents.

(Section 3.5 gives the background for this capability.)

6.2.3 The Edge Engine

The edge engine is basically a semantic engine, with some provisions for some pragmatic concerns (sect. 6.2.3.1.2). The edge engine is here viewed as providing for either of two modes for executing edges: either by directly executing the edge, or through relating two adjacent (edge connected) nodes.

6.2.3.1 New Node Production

Given one or more nodes all with edges leading to a ("new") node. To produce the document(s) corresponding to this latter node we need the documents of the former nodes. Once these have been established we can proceed. Given that we understand the overall principle for transforming the former documents into the new document, we now execute the edge by actually effecting, under developer control, but supported extensively by the edge semantic engine, the previously contemplated, and now (to be) detailed derivation. That is: the new node document arises by a process of deriving "old" documents into a "new" document. We see two aspects in this process: a semantic, and a pragmatic aspect.

6.2.3.1.1 The Semantic Part

We refrain at the present time from detailing this part, and refer instead to:

Program Derivations and Software Engineering
Ulrik Jørring
August 25, 1985
Carnegie-Mellon University / Dansk Datamatik Center

The edge semantic engine provides the capabilities described in section 4, pages 29-36, of the above-referenced report.

6.2.3.1.2 The Pragmatic Part

As for the node pragma engine we find a need for the edge pragma engine to likewise provide facilities for:

    .1   Recording Design Decisions
    .2   Tracking Requirements

6.2.3.2 Inter-Node Relations

Successor node documents can, on the other hand, also be developed "informally" by using the standard facilities and functions of the node engine, and then, only subsequently, relating the resulting document to its predecessor documents! This process of "relating back" can be viewed as consisting of four aspects.

6.2.3.2.1 Injection

Given abstract objects of a specification, or a high level design, and given some proposal for their object transformation into more concrete objects, a need arises for specifying an injection (also called a representation) relation which indicates how one might embed abstract objects in a concrete domain of objects.

The edge semantic engine provides functions for establishing and checking injection relations.

Establishing and checking such relations is also an engineering device for increasing confidence in the correctness of the chosen design in that it is an engineering process which affords another independent view of the specification/design relations.

6.2.3.2.2 Adequacy

Given abstract objects of a predecessor document, object and operation transformations design or imply more concrete objects. Adequacy here means establishing a predicate, ie. a relation which expresses that to each abstract object there corresponds a nonvoid (non-trivial) more concrete object which is capable of effecting ("realizing") described operations.

The edge semantic engine provide functions for establishing and checking such adequacy predicates.

6.2.3.2.3 Abstraction

Once a set of object and operation design choices have been chosen an engineering need exists for relating these back to the originating specification or high level design.

The edge semantic engine provides functions for establishing and checking such abstraction or retrieval functions, and for using them in proofs of correctness.

Again: the mere establishment of "correct" abstraction functions serves to increase confidence in the correctness of the design.

6.2.3.2.4 Correctness

In the case one can apply a-priori sound derivation rules to the production of designs from specifications there normally may be no need for proving correctness. But in the case where a design is informally "specified" without formal derivation rules having been established, there arises an a-posteriori proof obligation. The node/edge engine provides comprehensive functions for verifying such designs.

6.2.4 The Management Engine

The graph-node-edge engines provide facilities and functions for software engineers and programmers. The management engine provides facilities for management. These facilities shall serve the following management functions:

- *controlling resources*
- *performance monitoring*
- *productivity assessment*
- *quality assessment*
- *priority setting*
- *project authorisation*
- *procurement and contracts*
- *personnel management and recruiting, and*
- *security, recovery and auditing*

Besides the standard office automation facilities of:

- *electronic mail*
- *electronic conferencing*
- *advanced word-processing*
- *spread sheet / rolling plan analysis*
- *information storage, retrieval and analysis*
- *integrated graphics, and*
- *voice mail*

the management engine shall provide project graph related capabilities such as:

- *decision support system*
- *project task time management, and*
- *project meeting calendaring.*

The management engine will provide for the above facilities (which have been listed as they are seen in the traditional light of conventional management) through the functions described next (and which are listed in the new light of project graphs).

6.2.4.1 Project Graph Functions

Management must ultimately control and monitor all resource consumption (men, machines, money) on a per node and edge basis. Hence management must be directly involved in constructing and annotating project graphs. Project management annotations to nodes and edges focus on manpower consumption and computer usage estimation.

6.2.4.2 Resource Planning and Consumption Functions

Given a project graph, fully annotated by management wrt. resource

--Fig.6 *The Software Development Support System* --------------------

SOFTWARE DEVELOPMENT SYSTEM

- **PROGRAMMING PROCESSOR (The Graph Machine)** [6.2]
  - **GRAPH ENGINE** [6.2.1]
    - Input
    - Update
    - Output
  - **NODE ENGINE** [6.2.2]
    - *Syntax* [6.2.2.1]
      - Edit /Format - Journal [6.2.2.1.1-2]
    - *Semantics* [6.2.2.3]
      - Statics (type check, cross ref. ...) [6.2.2.3.1]
      - Dynamics (animation, prototyping) [6.2.2.3.2-3]
      - Validity (prove, verify) [6.2.2.3.4]
    - *Pragmatics* [6..2.2.2]
      - Annotate, Design Decisions [6.2.2.2.3-2]
      - Requirements Tracking [6.2.2.2.4]
      - Versions, Configurations [6.2.2..2.1]
    - *Inter Node* [6.2.3.2]
      - Injection/Representation [6.2.3.2.1]
      - Adequacy [6.2.3.2.2]
      - Abstraction/Retrieve [6.2.3.2.3]
  - **EDGE ENGINE** [6.2.3]
    - *New Node* [6.2.3.1]
      - Semantic [6..2.3.1.1]
        - Derivation : transform./enrich/refine
      - Pragmatic [6.2.3.1.2]
        - Design Decisions
        - Requirements
- **MANAGEMENT PROCESSOR** [6.2.4]
  - Graph Summarizer
  - Node/Edge Summarizer
  - Etc.
- **THEORY PROCESSOR** [6.2.5]
  - Tool Generator
  - Theory Inserter

consumption, on a per node and edge basis, and given minimum and maximum estimates of resource availability on, say a monthly basis, sufficiently well into a future, a variety of alternative sequentializations and parallellizations can be extracted. That is: (A) each node has attached to it estimates on how many people (min.max.) should work on it and for how long, similarly for edges. This is what it takes. Now (B) estimates are also provided on availability of resources, likewise on a monthly basis. Given the (acyclic) graph nature of the project graph, allowing in general for several node and edge activities to take place in parallel, one applies (A) to (B), and gets a set of ways in which to allcoate resources to nodes, edges and actual calendar months. This is planning. That is: control.

One particular such plan is chosen. The project starts. And we now go into monitoring progress. When actual progress is made one either exceeds estimates or not. In case of overspending adjustments need be made. Again the project graph, with a re-run of its allocation planning function, is the focus and basis for such project reorganisations. [See ref. [3].]

6.2.4.3 Miscellaneous Functions

The project graph with its original management annotations and with its attached, running reports on actual progress (resource consumption, completion status of tasks, etc.) can then be used to display (print out) various management reports. The idea is to represent the various kinds of information in a logical form such that relational queries can be uses, by management, to extract such things as time bar charts, rolling plans, etc.

6.2.5 The Theory Machine

(Perhaps this section should have been indexed 6.3 instead!)

The Graph Machine has a number of users: programmers (specifiers, designers, coders), software engineers, and managers. They all use functions of that machine. We have not, so far in this paper, committed the project graph idea to any one particular method. We believe that the software development system, SDS, as it reflects the project graph notion, is relatively independent of which particular method is embedded. Among the methods for which we can tailor the system we see: VDM, HDM, HOS, JSP/JSD, Yourdon/SA/SD, PROSPECTRA, MetaSoft, RAISE, etc.

So a first function of the Theory Machine is to take a "raw" (method-free) SDS, and tailor it to a particular method. The SDS notions of documents, annotations, etc., apply to any, or at least most methods. The facets which are method dependent are:

> the syntax of the way in which specification
> and design documents are written, and
>
> their semantics.

Hence the Theory Machine will have facilities for personalizing syntax and semantics.

As a method-tailored SDS evolves additional, new syntax and semantics has to be added.

6.2.5.1 The Syntax Theory Modifier

Thus the Theory Machine will include:

      lexical scanner generators,
      syntax parser generators, and
      attribute grammar evaluator generators

to allow for insertion of new syntaxes: the tokenisation, parsing and type (etc.) checking of input.

6.2.5.2 The Semantics Theory Modifier

And the Theory Machine will include:

      rewrite rule systems and their generators,
      unifiers,
      theorem provers, proof editors and verifiers

to allow for insertion of new meanings: interpreters (symbolic and otherwise).

7. TOWARDS A THEORY OF SOFTWARE DEVELOPMENT

We are moving towards a close. What have we achieved? We claim to have basically reformulated ideas that have always been part of the folklore of computation, by viewing them in one (of several) kind(s) of context, namely that of project graphs. Project graphs could certainly be half-dismissed as mere PERT Diagrams. We believe, however, that that would be a mistake. PERT Diagrams have only a management and only partially an engineering semantics. The System Development System, Project Graphs and their associated documents, annotations, etc., contain, in a sense all the meaning of what software development entails. One has, so to speak, wrapped the entire spectrum: from theoretical computer science, via computing science (programming methodology) and software engineering to management, into one notion: that of Project Graphs as here explained.

It is said by leading programming methodologists, software engineers and software development managers, that although all software development projects contain innumerable problems, there is only one commonality among them: that the project people: programmers, engineers, and managers, do not know what they are doing. With the common notion of project graphs, with the insistance that they be developed prior to any actual development, and that they be strictly followed, by all, during development, and with the belief that the notion of Project Graph is indeed a most natural notion, we believe to have offered a solution to exactly that common problem!

The Dogma is:

Do not pursue any software development before you have a clear understanding of what its project graph is, to a depth more detailed than that of the example of section 1.

8. ACKNOWLEDGEMENTS

I thank my colleagues, at Dansk Datamatik Center, at my Computer Science Department of the Technical University of Denmark, and at STC (STL Harlow) UK, for listening to my ideas.

I thank B.T. Denvir, R.C. Shaw, M.I. Jackson, A. Munk Madsen, for extensively commenting on them.

And I thank my students, N. Nørgaard, I. Lysgaard Andersen, S.Aa. Fly Hansen, M. Jess, M. Frendorf, and S.M.Lynenskjold for working on them.

9. REFERENCES

[Bj|rner 77]   *Programming Languages: Formal Development of Interpreters and Compilers*,   in: Proceedings,   Intl.   Comp.   Symp., ICS'77, North Holland Publ.,   pp. 1-21.

[Bj|rner 80]   -- & Ole N. Oest: *The DDC Ada Compiler Development Project*, in: "Towards a Formal Description of Ada", Springer Lecture Notes, vol. 98, Nov. 1980, pp. 1-19.

[Bj|rner 82]   -- & S.Prehn: *Software Engineering Aspects of VDM*,   in: "Theory and Practice of Software Technology",   Proc., Intl. Seminar, Capri, May 1982, (ed. Ferrari, Bolognani, Goguen) North-Holland Publ., 1983.

[Scott 83]   -- & W.Scherlis: *Inferential Programming*, in: Proc. IFIP Congress '83, North-Holland Publ., 1983.

[Sintzoff 84]   *Meta-Programs*, in: Proc., Workshop on "Combining Specification", Nyborg, Denmark, May 1984, (eds. Bj|rner & Prehn) Springer NATO Advanced Systems Institute Series, forthcoming 1986.

10. BIBLIOGRAPHY

[1]   Dines Bjørner: *DiProGS: A Distributed Project Graph System*, Dept. of Comp. Sci., Techn. Univ. of Denmark, 20. April 1986, 22 pages.

[2]   N.Nørgaard, I.Lysgaard Andersen, S.Aa Fly Hansen, M. Jess, and M. Frendorf: *A Formal Specification of Project Graphs*, Dept. of Comp. Sci., Techn. Univ. of Denmark, July 1986, 61 pages.

[3]   S.Lynenskjold (et al.): *A Project Graph Resource Allocator*, Dept. of Comp. Sci., Techn. Univ. of Denmark, Fall 1986.

Software Development based on Formal Methods*

C.B. Jones
Department of Computer Science
University of Manchester
M13 9PL, United Kingdom

October 19, 1986

Abstract

Formal methods employ mathematical notation to record specifications and use mathematical reasoning to justify designs with respect to such specifications. One avenue of formal methods research is known as the Vienna Development Method. *VDM* has been used on programming language and non-language applications. In this paper, programming languages and their compilers are ignored; the focus is on the specification and verification of programs. The discussion focusses on the approach to specification and the reliance on proof obligations in design steps of data reification and operation decomposition.

1 Introduction

The term "Formal Methods" applies to the use of mathematical notation in the specification, and the use of such specifications as a basis for the verified design, of computer systems. This paper outlines one such formal method: *VDM* ("Vienna Development Method") as applied to program development—for a fuller account see [22]. (*VDM* has also been used for the description of programming language semantics and the development of compilers—see [4]).

There are three more or less distinct approaches to the formal development of programs. Each approach starts with a formal statement of the required function of the program and uses formally provable steps to link the final program to the specification. The approaches and their emphases are:

- specification/design/verification: the specification is written in a distinct specification language but the design (and eventual code) are written in a normal implementation language; correctness of the design is established by discharging defined proof obligations. *VDM* uses this approach.

---

*This paper is an extended version of "Specifications and Programs" which was presented at, and will appear in the proceedings of, the "Third Autumn School of the Polish Informatics Society, Mragowo, Poland, 1986.

- transformation: the "specification" is given as an executable function but one where clarity is considered to be paramount (the execution of the function is likely to be extremely inefficient); an acceptable implementation is created by a series of syntactic transformations (with some of which are associated the obligation to prove applicability). The best known example of this approach is the CIP project (see [8]).

- constructive mathematics: here the specification is taken as the statement of a theorem from whose constructive proof, can be extracted a program (cf. [9]).

VDM uses mainly model-oriented specifications. These are contrasted to property-oriented specifications in the next section and the respective rôles of the two methods are discussed. Operations are specified by post-conditions of initial and final states as well as (separate) pre-conditions.

Section 3 shows how model-oriented specifications are useful in capturing the architecture of a system. Examples are used in Sections 4 and 5 to illustrate the systematic design aspects of *VDM*. (The examples are given here without formal proofs.)

2 Data Types

There are two schools of data type specification. They are referred to here as the *property-oriented* and *model-oriented* approaches. Both approaches have their uses and, if employed appropriately, complement each other. Suppose that it is wished to describe a basic data type like finite sets of natural numbers. The signatures of the operators could be:

$empty: \rightarrow Set_{\mathbf{N}}$
$add: \mathbf{N} \times Set_{\mathbf{N}} \rightarrow Set_{\mathbf{N}}$
$is\text{-}empty: Set_{\mathbf{N}} \rightarrow \mathbf{B}$
$is\text{-}memb: \mathbf{N} \times Set_{\mathbf{N}} \rightarrow \mathbf{B}$

With these operators, it is possible to generate terms like:

$add(3, add(5, empty()))$

or propositions like:

$is\text{-}memb(5, (add(3, add(5, empty()))))$

It would clearly be more natural to introduce the standard infix operator symbol ($\in$) for *is-memb* and a minor extension to the way of presenting signatures would permit this. But the signature represents only the syntactic part of the description of the data type. The essential difference between property and model descriptions comes in the way that the semantics is presented. In a property-oriented description, the meaning of the operators is fixed by equations[1]. The key to the creation of these equations is to recognise—in the case in hand—that all finite sets can be *generated* by the operators *empty* and *add*. It is then straightforward to characterize those operators which deliver values in *externally visible* types (i.e. Natural numbers and Booleans) in terms of the generators. For example:

$is\text{-}empty(empty()) = \mathsf{true}$
$is\text{-}empty(add(i, s)) = \mathsf{false}$

captures, in some way, obvious properties of the *is-empty* operator. Similarly, the *is-memb* operator can be described by the equations:

---

[1]The use of such algebraic equations gives rise to the more commonly used names for this approach: *equational specifications* or *algebraic presentations* or even *algebraic specifications*. This last is somewhat of a misnomer since it is as acceptable to present a model of rational numbers as it is to give the axioms of natural numbers in an algebraic text (cf. [27]).

$$is\text{-}memb(i, empty()) = \mathsf{false}$$
$$is\text{-}memb(i, add(j, s)) = (i = j \lor is\text{-}memb(i, s))$$

This example is very simple but it makes it possible to discuss the strengths and weaknesses of property-oriented specifications. The most obvious advantage of such a description is that it is presented without reference to any underlying, or pre-defined, data type. In fact, the rôle of a model is provided by the valid terms (*word algebra*) which can be built from the generators[2]. A more subtle advantage derives from the fact that the whole concept is built on a branch of mathematics (i.e. Algebra) where notions relevant to data types have been studied. In particular *signatures*, *sorts*, *equations* and *models* are all of interest. The generalisation from a specific data type like Set_N to a set type which is parameterized by the type of its elements is most easily studied in the property-oriented approach.

The choice between a property and a model specification should be made on pragmatic considerations. There are, however, some technical difficulties with property-oriented descriptions which should be understood. In the Set_N example, all of the operators are *total*; had the task been to define sequences of natural numbers, the operator to take the first element of a list (hd) would have been *partial*. Partial operators arise very often in computing and a treatment which fits the way in which they are used is essential. The first major approach to the handling of *error algebras* (cf. [12]) was less than satisfactory; more recent work (e.g. [6]) fits more closely the way in which partial operators are used.

A second area of difficulty is the question of *interpretations* of such equations[3]. The choice between *initial*, *loose* or *final* interpretations are too technical to pursue in detail here. Section 9.2 of [22] gives a brief overview of how either extra operators or equations are necessary in the final and initial approaches respectively in order to ensure the appropriate identifications. For a fuller treatment, the reader is referred to a textbook such as [3] or [11].

A far deeper problem comes from the fundamental limit on the expressive power of a specification by properties. It has long been known that certain data types cannot be characterized by finite sets of equations. This gives rise to the need for, so-called, *hidden functions*. The relationship between these functions and a model is an interesting topic for research. For now, it is more important to observe that the presence of such hidden functions weakens the main advantage of a property-oriented specification: the ideal that a data type can be understood solely in terms of its operators (or functions) and their relationship is clearly unattainable if new functions have to be introduced to describe the inter-relationship.

It is now time to turn to the pragmatic questions which are likely to govern the choice between a property and a model based description of a data type. A distinction can be made between data types like Set_N which possess no obvious state and those like a database where the concept of a state affecting, and being changed by, the execution of operations[4] is pervasive. In fact, even with the example of a *Stack*—almost the standard example of a data type specification—there is a natural place for a state. It is possible to disguise this fact by presenting a signature of the form:

$$empty: \rightarrow Stack$$
$$push: X \times Stack \rightarrow Stack$$
$$top: Stack \rightarrow X$$
$$remove: Stack \rightarrow Stack$$
$$is\text{-}empty: Stack \rightarrow \mathbf{B}$$

---

[2]Unfortunately the generators do not always provide a convenient model (cf. FIFO queues); nor is it always the case that a unique term exists for each distinct element of the model.

[3]This is also connected with the question of what happens when a convenient set of generator operators is not present.

[4]The term *operations* is used in preference to "operators" in order to emphasize the rôle of side-effects.

But this separates the two parts (*top* and *remove*) of what is naturally a single *POP* operation which changes the state by side effect and delivers the required value as a result. There is no basic reason why property based descriptions could not be extended to handle signatures with more than one result (see, for example, [28][5]). It does, however, remove some of the elegance with which the defining equations can be presented.

The alternative, model-oriented, specification style handles the operations separately. Each operation is characterized by pre- and post-conditions in which there is no difficulty in handling state-like objects. The obvious danger presented by basing a specification on a model is that of "over-specification". This problem has been characterized in [17] and [22] as *implementation bias*: a test is given there which establishes that the underlying state exhibits no bias.

A series of operations are each specified with respect to a state; the state is constructed from combinations of known types. Picking up the example of a stack, the underlying state might be defined in terms of sequences of X. The *POP* operation could then be specified:

POP () $r:X$

ext wr st : seq of X

pre $st \neq [\,]$

post $r = $ hd $\overleftarrow{st} \wedge st = $ tl $\overleftarrow{st}$

The ext clause identifies the non-local objects to which the operation has access. In this case, there is only one variable to be considered since the state is so simple. In larger examples, listing only those variables required goes some way to solving the so-called "frame problem". Furthermore, distinguishing between read-only access (rd) and read-write access (wr) can also clarify the potential effect of an operation. The pre-condition is a predicate of a state and can be used to limit the cases in which the operation has to be applicable (here, the implementor of *POP* is invited to ignore states in which the initial st is an empty sequence). The post-condition is a predicate of two states: it specifies the relationship required between the states before and after execution of the operation. Here, then, it is necessary to distinguish two values of the same (external) variables. There are many possible conventions for doing this; in [22] the values in the old state are decorated with a hook (e.g. $\overleftarrow{st}$).

Such an operation specification could be translated into a more functional notation:

POP: seq of $X \rightarrow$ seq of $X \times X$

This paper uses the more schematic style of *VDM* operation specification throughout. The decision to separate the pre-condition in an operation specification is made on pragmatic grounds. Partial operations are very common in computing and the pre-condition focuses attention on the assumptions. The full power of the post-conditions becomes apparent on more complex examples. For now, the advantages are simply listed:

- the ability to specify non-deterministic (and thus under-determined) operations;

- results can often be conveniently specified by conjunctions of properties—this makes it far easier to specify, than it is to create, the result;

- the use of negation has a similar effect;

- it is often easy to specify an operation in terms of some inverse.

---

[5]In the presence of non-determinism, the attempt to separate operations into functions which only deliver one result is invalid.

Both partiality and non-determinism[6] present problems in property-oriented specification techniques[7].

One disadvantage of model-oriented specifications by pre- and post-conditions is that it is possible to specify an operation which is unimplementable (e.g. producing an even prime number greater than 10). This gives rise to the first of many *proof obligations* which are an inherent part of *VDM*. An operation (e.g. *POP*) is *implementable* only if:

$$\forall \overleftarrow{st} \in \text{seq of } X \cdot pre\text{-}POP(\overleftarrow{st}) \Rightarrow$$
$$\exists st \in \text{seq of } X, r \in X \cdot post\text{-}POP(\overleftarrow{st}, st, r)$$

These particular proof obligations are not normally subject to formal proof but do provide a convenient reminder that type information, pre-condition and post-condition all combine to govern whether an operation is implementable.

3   Abstract Models

It is pointed out above that—in model-oriented specifications—operations can be considered separately. In this section, it is shown that the structure of a state can be used to study the *architecture* of a system before any operations are considered.

Suppose that the task is to design and specify a file system. For this purpose, it is possible to ignore the internal structure of a *File* (it might be a sequence of bytes). In order to access the files, they are given names (*Name*). A trivial file system might be defined on states which contain only:

$$Trivial = \text{map } Name \text{ to } File$$

(The map objects, like sets and sequences, are basic ways of building composite objects in *VDM*.) On such a state, it would be possible to define operations to *CREATE*, *DELETE* and *COPY* files. But it is also necessary to observe what cannot be done. From the properties of maps, it follows that no two files can have the same name. Thus, in this trivial system, there is no support for different users to be given different name spaces. The system is not rich enough and this can be seen from the state even before operation specifications are written.

Separate name spaces could be created by nested directories. The state for such an enriched system could be defined:

$$Nestedfs = Directory$$

$$Directory = \text{map } Name \text{ to } Node$$

$$Node = Directory \cup File$$

Such a system would allow different users to employ the same names. The directory structure is, in many respects, similar to that of Unix-like systems: the *Node* concept reflects the way in which files and directories can occur in the same directory.

Here again, it would be possible to define operations on *Nestedfs*. But it is also wise to check what cannot be done in this state. It is not possible to share the same *File* via different path names (sequences of names). If it were wished to permit a change via one path name to affect access via other path names, the state would again have to be extended.

---

[6]An interesting approach of *partial interpretations* is described in [6]. This handles under-determined but not non-deterministic functions. The need for this latter, even in a deterministic implementation, comes from the change of meaning of equality at different levels of abstraction.

[7]See, however, [28].

There is a relatively standard way of introducing such sharing patterns into specifications. An intermediate link (in this case, *Fid*) is introduced[8]:

Sharedfs ::   *root* : *Directory*
        filem : map *Fid* to *File*

Directory = map *Name* to *Node*

Node = *Directory* ∪ *Fid*

A single file can now be shared as in the following object:

$mk\text{-}Sharedfs(\{id_1 \mapsto fid_1, id_2 \mapsto \{id_1 \mapsto fid_2, id_2 \mapsto fid_1\}\},$
$\qquad\qquad \{fid_1 \mapsto file_a, fid_2 \mapsto file_b\})$

Having decided that such a state supports the overall level of function required, it is now possible to define operations on this state. One operation can be defined to show the contents of a directory:

Dirstatus = map *Name* to {*FILE, DIR*}

SHOW () r: *Dirstatus*
ext rd *d* : *Directory*
post $r = \{nm \mapsto (\text{if } d(nm) \in Directory \text{ then } DIR \text{ else } FILE) \mid nm \in \text{dom } d\}$

Another operation might create a new directory within an existing one:

MKDIR (*n*: *Name*)
ext wr *d* : *Directory*
pre $n \notin \text{dom } d$
post $d = \overleftarrow{d} \cup \{n \mapsto \{\}\}$

An operation to insert a new file might be defined:

MKFILE (*n*: *Name*, *f*: *File*)
ext wr *d* : *Directory*
    wr *fm* : map *Fid* to *File*
pre $n \notin \text{dom } d$
post $\exists fid \in Fid \cdot fid \notin \text{dom } \overleftarrow{fm} \wedge d = \overleftarrow{d} \cup \{n \mapsto fid\} \wedge fm = \overleftarrow{fm} \cup \{fid \mapsto f\}$

The interested reader should be able to define other operations (e.g. for deletion) at the level of a *Directory*.

One of the reasons that the work on property-oriented specifications has been important is that convenient ways of structuring specifications have been studied. Section 7.4 of [22] employs a technique for *promoting* operations from one data type to another by *operation quotation*. This technique can be used to apply operations on a single directory to the whole directory structure. For this purpose, the state might be extended with components which—for example—contain the current path. This directory example shows how the architecture of a system can be studied via its state. Other examples in [22] show how a system like virtual storage can be studied at

---

[8]The *VDM* composite object notation is somewhat like Pascal records. In this case:

$Sharedfs = \{mk\text{-}Sharedfs(root, filem) \mid root \in Directory \wedge filem \in \text{map } Fid \text{ to } File\}$

different levels of abstraction in order to bring out different concepts at each level. Section 8.3 of the same reference shows how the same ideas can be used to describe interfaces. In [4], similar ideas are applied to problems of programming languages.

4 Development by Data Reification

A specification of a single operation contains a pre- and a post-condition. As pointed out above, this makes it possible to define partial and non-deterministic operations. Intuitively, it should be acceptable for an implementation to terminate on more inputs than are required. (i.e. have a bigger domain—or be defined on more input values) or to produce some subset of the permitted answers for any required input (i.e. be more determined). An implementation is written in some implementation language and it is therefore necessary to have some common way of discussing the meaning of both specifications and of programs. One convenient model is to define both in terms of the set of states over which termination is required and the meaning relation which defines the possible results:

$$(S, R)$$

where S is a subset of the set of states, say Σ, and R is a relation on Σ.

It is clear that a given *pre*, *post* pair can be translated into this form by:

$$(\{\sigma \in \Sigma \mid pre(\sigma)\}, \{(\overleftarrow{\sigma}, \sigma) \in \Sigma \times \Sigma \mid post(\overleftarrow{\sigma}, \sigma)\})$$

The ideas of denotational semantics can be used to express programming language constructs in terms of the same model (see, for example, [19]).

In terms of the set/relation pairs, it is possible to define precisely the notion of *satisfaction* (sat), which is described intuitively above[9]:

$$(S_1, R_1) \text{ sat } (S_2, R_2) \quad \Leftrightarrow \quad S_2 \subseteq S_1 \wedge S_2 \triangleleft R_1 \subseteq R_2$$

This sat relation provides the basis against which steps of development must be shown to be correct.

There are two things which make specifications far shorter than programs: they use data objects (e.g. maps) which are not present in most programming languages and they use post-conditions which do not show how to compute a result. Both of these specification "tricks" have to be removed in the design of a program. The realization of the data object is handled in steps of *data reification*; the development of control constructs to satisfy post-conditions is handled by steps of *operation decomposition*. In the development of any significant system, development will take place in many steps. Experience with *VDM* suggests that the early, or high-level, design stages concern data reification and the later, low-level, steps concern operation decomposition. Both of these sorts of steps are illustrated below and the *proof obligations* necessary to establish correctness of such steps are explained in terms of examples. These proof obligations are sequents which must be true for the design step to satisfy its specification. The required proofs can be conducted at an appropriate level of formality.

In the extreme, these proof obligations could be compared to the output of *verification condition generators* (VCG) (cf. [5]). The VCG approach has received considerable criticism because of the difficulty of relating the created logical expressions to the original programming task. This author's own experience both with hand proofs and with Jim King's *EFFIGY* system (cf. [24]) has confirmed the validity of this criticism. The attempt to prove a complete program correct using VCG's is like trying to solve equations in large numbers of variables—unfortunately, failure to find a proof corresponds to the lack of a solution and the ensuing hunt

---

[9]$S \triangleleft R \quad \triangleq \quad \{(z, z') \in R \mid z \in S\}$

for alternative assertions is very tedious. How does *VDM* avoid this problem? The approach is to use the steps of development themselves as a way of decomposing the correctness argument. Well-chosen steps provide an informal proof outline of the type used by mathematicians. This reduces the need for formal proof. If it is decided to construct a formal proof, such a proof is likely to be relatively simple. Perhaps most importantly, any errors are relatively easy to locate. It has been argued elsewhere (e.g. [21]) that this approach can increase the productivity of the development process by locating—soon after insertion—any errors which are made in the early stages of design.

The remainder of this section illustrates data reification steps via an example. The specification is a simple one which involves the storage of a set of objects of some given type—say, X. The initial state is the empty set:

$$s_0 = \{\,\}$$

An operation to add a new element to a set might be specified:

ADD $(e\colon X)$
ext wr s : set of X
pre $e \notin s$
post $s = \overleftarrow{s} \cup \{e\}$

To delete an element:

DELETE $(e\colon X)$
ext wr s : set of X
pre $e \in s$
post $s = \overleftarrow{s} - \{e\}$

An operation to test whether an element is present can be specified:

ISPRESENT $(e\colon X)\ r\colon \mathbf{B}$
ext rd s : set of X
post $r \Leftrightarrow e \in s$

This specification looks trivial because the chosen state objects match the problem exactly. Strictly, the implementability (see above) proof obligation should be considered for each operation. Here, the result follows immediately from totality of the set operators.

One way of building a representation for sets, which permits efficient access, is to store the elements in a *binary tree*. Such trees:

- have two (possibly nil) branches and an element (of X) at each node;

- are arranged so that all elements in the left branch of a node are less than[10] (and all elements in the right branch are greater than) the element in the node.

Formally, the *Setrep* data structure is defined:

$Setrep = [Node]$

$Node ::\quad lt\ :\ Setrep$
$\qquad\qquad mv\ :\ X$
$\qquad\qquad rt\ :\ Setrep$

---

[10]For brevity, the ordering relation is written as $<$.

where[11]:

$$inv\text{-}Node(mk\text{-}Node(lt, mv, rv)) \quad \triangle$$
$$(\forall lv \in retrns(lt) \cdot lv < mv) \wedge (\forall rv \in retrns(rt) \cdot mv < rv)$$

$retrns : Setrep \rightarrow$ set of X

$retrns(sr) \quad \triangle$ cases sr of
$$nil \rightarrow \{\}$$
$$mk\text{-}Node(lt, mv, rt) \rightarrow retrns(lt) \cup \{mv\} \cup retrns(rt)$$
    end

Defining such a representation is the first step in a *data reification*. The next step is to define the relationship between the abstract structure and the representation. For each abstract object (set of X) there are many possible representations (*Setrep*). This is a typical situation and therefore it is convenient to define the relationship between the abstraction and the representation by a function from the latter to the former (called, in *VDM*, a *retrieve function*). In this simple case, the function *retrns* is exactly what is needed. Notice that this function is total over *Setrep*. The next step in data reification is to establish the proof obligation known as *adequacy*. This requires that there is at least one representation for each element of the abstract states:

$$\forall s \in \text{set of } X \cdot \exists sr \in Setrep \cdot retrns(sr) = s$$

This could be proved formally by induction on the set generators; here an informal argument would suffice.

Having established the basic properties of the data representation, it is necessary to define each of the operations on *Setrep*. Thus, for example:

$ADD_1 \ (e{:}X)$
ext wr sr : $Setrep$
pre $e \in retrns(sr)$
post $retrns(sr) = retrns(\overleftarrow{sr}) \cup \{e\}$

might be given as the specification of an operation which is intended to mirror the behaviour of *ADD*. Notice that this post-condition is non-deterministic in that there are—except in trivial cases—many possible results which would be acceptable. This illustrates the way in which non-determinism can be used to structure a design. Even if the final program would be deterministic, this post-condition makes it possible to record and justify the design decision to use binary trees whilst postponing the decision about the tree balancing algorithm.

For the sake of simplicity, a more definite post-condition is used here (the idea of *operation quotation* is employed for illustration):

$ADD_1 \ (e{:}X)$
ext wr sr : $Setrep$
pre $e \notin retrns(sr)$

---

[11]For efficiency, such trees should also be *balanced*; for brevity, this requirement is not treated formally here.

post $\overleftarrow{sr} = $ nll $\wedge sr = mk\text{-}Node(nll, e, nll) \vee$
    $\overleftarrow{sr} \in Node \wedge$
    let $mk\text{-}Node(\overleftarrow{lt}, mv, \overleftarrow{rt}) = \overleftarrow{sr}$ in
    $e < mv \wedge$
        $\exists lt \in Setrep \cdot post\text{-}ADD_1(e, \overleftarrow{lt}, lt) \wedge sr = mk\text{-}Node(lt, mv, \overleftarrow{rt}) \vee$
    $e > mv \wedge$
        $\exists rt \in Setrep \cdot post\text{-}ADD_1(e, \overleftarrow{rt}, rt) \wedge sr = mk\text{-}Node(\overleftarrow{lt}, mv, rt)$

This records the essential recursion which could be used in ADD_1 but preserves the requirement that post-conditions are simply predicates (it is not possible to invoke an operation from within a predicate).

At this level of design, the obligation to prove implementability is non-trivial. That this algorithm yields an object which satisfies the data type invariant ($inv\text{-}Node$), should be proved. Thus:

$$\forall \overleftarrow{sr} \in Setrep, e \in X \cdot pre\text{-}ADD_1(e, \overleftarrow{sr}) \Rightarrow \exists sr \in Setrep \cdot post\text{-}ADD_1(e, \overleftarrow{sr}, sr)$$

Such proofs are shown formally in [22].

The model of the $DELETE$ operation can be specified:

$DELETE_1$ $(e\colon X)$
ext wr sr : $Setrep$
pre $e \in retrns(sr)$
post $retrns(sr) = retrns(\overleftarrow{sr}) - \{e\}$

Here, the removal of the non-determinacy is less easy and the design of a specific recursive algorithm is left as an exercise.

The model of $ISPRESENT$ can be specified as:

$ISPRESENT_1$ $(e\colon X)$ $r\colon \mathbf{B}$
ext rd sr : $Setrep$
post $\overleftarrow{sr} = $ nll $\wedge \neg r \vee$
    $\overleftarrow{sr} \in Node \wedge$
    let $mk\text{-}Node(\overleftarrow{lt}, mv, \overleftarrow{rt}) = \overleftarrow{sr}$ in
    $e = mv \vee$
    $e < mv \wedge post\text{-}ISPRESENT_1(e, \overleftarrow{lt}, r) \vee$
    $e > mv \wedge post\text{-}ISPRESENT_1(e, \overleftarrow{rt}, r)$

Each of these models must be justified with respect to the more abstract specification. The proof obligations for operation data reification are:

$$\forall sr \in Setrep \cdot pre\text{-}OP(retrns(sr)) \Rightarrow pre\text{-}OP_1(sr)$$

for the domain part of the rule; and:

$$\forall \overleftarrow{sr}, sr \in Setrep \cdot pre\text{-}OP(retrns(\overleftarrow{sr})) \wedge post\text{-}OP_1(\overleftarrow{sr}, sr) \Rightarrow$$
$$post\text{-}OP(retrns(\overleftarrow{sr}), retrns(sr))$$

for the result part. (These rules require obvious extensions to cope with arguments and results.)

In general, since data reification steps come earlier in the design process, such proofs should be undertaken rather formally. Illustrations of such proof are given in Section 8.3 of [22].

These tree-like data structures show the essential idea behind binary trees. The data structures are not, themselves, representable in most programming languages because of their recursive nature. In, for example, Pascal it would be necessary to represent such trees via pointers and objects allocated on the heap. Still within *VDM* notation, this could be described as:

$root = [Ptr]$

$Heap = \text{map. } Ptr \text{ to } Node_2$

$Node_2 :: \quad lt \ : \ [Ptr]$
$\qquad\qquad mv \ : \ X$
$\qquad\qquad rt \ : \ [Ptr]$

where:

$inv\text{-}Heap : [Ptr] \times Heap \to \mathbf{B}$

$inv\text{-}Heap(p, m) \quad \underline{\triangle} \quad \ldots$

Following the pattern set above, the next step is to write a retrieve function:

$retrsr : [Ptr] \times Heap \to Setrep$

$retrsr(p, h) \quad \underline{\triangle} \quad \text{if } p = \text{nil}$
$\qquad\qquad\qquad\qquad \text{then nil}$
$\qquad\qquad\qquad\qquad \text{else let } mk\text{-}Node_2(lt, mv, rt) = h(p) \text{ in}$
$\qquad\qquad\qquad\qquad\qquad mk\text{-}Node(retrsr(lt, h), mv, retrsr(rt, h))$

This stage of reification brings the data structures close to those of potential implementation languages. The specification are, however, still not executable: pre- and post-conditions must be decomposed into sequences of statements. In *VDM* this sort of development step is known as *operation decomposition*. The example above is taken in [22] through steps of decomposition; here, space dictates the use of smaller examples.

5 Development by Operation Decomposition

The process of operation decomposition also gives rise to proof obligations. There follows an explanation of the proof obligations and a style in which programs can be annotated with their correctness arguments. Ways in which the proof ideas can be used in the development of programs are discussed in [23].

This section presents proof rules only for some simple programming language constructs. The general form of these rules is similar to those of logic. Here, the conditions under which a rule can be applied require that certain properties hold for sub-operations and the conclusions are that (other) properties hold for combinations of the sub-operations. The proof rules facilitate proofs that pieces of program satisfy specifications. Thus it can be shown that:

If $i < 0$ then $i, j := -i, -j$ else skip

satisfies the specification:

$MAKEPOS$ ()
ext wr i : $\mathbf{Z}$,
$\quad$ wr j : $\mathbf{Z}$
pre true
post $0 \le i \land i * j = \overleftarrow{i} * \overleftarrow{j}$

For small examples, it is convenient to record the specification and its implementation together, thus:

MAKEPOS
ext wr i: **Z**, wr j: **Z**
pre true
   If $i < 0$ then $i, j := -i, -j$ else skip

post $0 \leq i \wedge i * j = \overleftarrow{i} * \overleftarrow{j}$

Such an annotated program is written when the code has been shown to satisfy the specification. The name of the operation and the externals line are sometimes omitted when they are clear from context.

A natural extension of this style is to write specifications for sub-operations—rather than their code; see Figure 1. This can be read as saying that the composition of two sub-operations *MAKEPOS* and *POSMULT* would satisfy the specification of *MULT*. The sub-operations are not (yet) coded—rather, their specifications are given[12]. The proof rule for sequential execution is discussed below.

MULT
ext wr i, j, m: **Z**
pre true
   MAKEPOS
   ext wr i, j: **Z**
   pre true
   post $i \geq 0 \wedge i * j = \overleftarrow{i} * \overleftarrow{j}$
   ;
   POSMULT
   pre $i \geq 0$
   post $m = \overleftarrow{i} * \overleftarrow{j}$
post $m = \overleftarrow{i} * \overleftarrow{j}$

Figure 1: Example of Specifications in Place of Code

The proof rules are somewhat similar to the deduction rules for logic. Broadly, there is one proof rule for each language construct. In order to present the proof rules in a compact way, the pre- and post-conditions are written in braces before and after the piece of code to which they relate—thus:

$\{pre\} S \{post\}$

It is sometimes necessary to use information from a pre-condition in the post-condition. Decorating P with a hook to denote a logical expression which is the same as P except that all free variables are decorated with a hook, the relevant proof rule is:

$$\frac{\{P\} S \{R\}}{\{P\} S \{\overleftarrow{P} \wedge R\}}$$

---

[12] A design can be presented as a combination of (specified) sub-problems. A *compositional* development method permits the verification of a design in terms of the specification of its (syntactic) sub-programs. Thus, one step of development is independent of subsequent steps in the sense that any implementation of a sub-program can be used to form the implementation of the specification which gave rise to the sub-specification. In a non-compositional development method, the correctness of one step of development might depend on the subsequent development of the sub-programs. Compositional development methods are not too difficult to find for sequential programs. For programs which permit interference of parallel processes, the challenge is much greater. Some work in the *VDM* framework is reported in [20]; more recent work on Temporal Logic is described in [2].

Thus, for example, from:

$$\{fn = 1\} FACTB \{fn = \overleftarrow{fn} * \overleftarrow{n}\,!\}$$

it follows that:

$$\{fn = 1\} FACTB \{fn = \overleftarrow{fn} * \overleftarrow{n}\,! \wedge \overleftarrow{fn} = 1\}$$

and thus:

$$\{fn = 1\} FACTB \{fn = \overleftarrow{n}\,!\}$$

Notice that the hooking of P (and thus its free variables) is crucial—it is not true that, in the final state:

$$fn = 1$$

The most basic way of combining two operations is to execute them in sequence. It would be reasonable to expect that the first operation must leave the state so that the pre-condition of the second operation is satisfied. In order to write this, a distinction must be made between the relational and single-state properties guaranteed by the first statement. The names of the truth-valued functions have been chosen as a reminder of the distinction between:

$$P: \Sigma \to \mathbf{B}$$
$$R: \Sigma \times \Sigma \to \mathbf{B}$$

Writing:

$$R_1 \mid R_2$$

for:

$$\exists \sigma_i \cdot R_1(\overleftarrow{\sigma}, \sigma_i) \wedge R_2(\sigma_i, \sigma)$$

the sequence rule is:

$$\frac{\{P_1\}S_1\{P_2 \wedge R_1\},\ \{P_2\}S_2\{R_2\}}{\{P_1\}S_1; S_2\{R_1 \mid R_2\}}$$

The predicate P_2 can be seen as the designer's choice of interface between S_1 and S_2 whereas R_1 and R_2 fix the functionality of the two components.

In Figure 1, the first conjunct of *post-MAKEPOS* is also contained in *pre-POSMULT*: this defines the interface between the two (as yet to be coded) components. The condition given as *post-POSMULT* is the same as *post-MULT* but the position of the former shows that its hooked variables refer to values which will arise between the execution of *MAKEPOS* and *POSMULT*. The second conjunct of *post-MAKEPOS* is the simplest expression (for R_1 in the rule) which ensures that *post-MULT* is satisfied—in detail, *post-MAKEPOS* can be written as:

$$\textit{post-MAKEPOS}(\overleftarrow{i}, \overleftarrow{j}, \overleftarrow{m}, i, j, m) \quad \triangle \quad i * j = \overleftarrow{i} * \overleftarrow{j} \wedge m = \overleftarrow{m}$$

(The non-appearance of m in the external clause of *MAKEPOS* justifies the second conjunct.) Also:

$$\textit{post-POSMULT}(\overleftarrow{i}, \overleftarrow{j}, \overleftarrow{m}, i, j, m) \quad \triangle \quad m = \overleftarrow{i} * \overleftarrow{j}$$

Thus:

$$\textit{post-MAKEPOS} \mid \textit{post-POSMULT}$$

becomes:

$$\exists i_i, j_i, m_i \cdot i_i * j_i = \overleftarrow{i} * \overleftarrow{j} \wedge m_i = \overleftarrow{m} \wedge m = i_i * j_i$$

from which:

$$m = \overleftarrow{i} * \overleftarrow{j}$$

follows.

In practice, it is not normally necessary to proceed formally with such proofs. It becomes rather easy to check an annotated text like that for *MULT*. The only care required is the association of the hooked variables with the values at the beginning of the appropriate operation. A good visual check is given by the nesting. (The generalization to a sequence of more than two statements is straightforward.)

The proof rule for conditional statements is:

$$\frac{\{P \wedge B\} TH \{R\}, \ \{P \wedge \neg B\} EL \{R\}}{\{P\} \text{ if } B \text{ then } TH \text{ else } EL \{R\}}$$

Looking at this proof rule, it would appear that the designer has little freedom of choice other than the selection of cases. Consideration of even a simple case—again taken from the multiplication example—shows how the designer's freedom actually arises:

MAKEPOS
ext wr $i, j : \mathbf{Z}$
pre true
If $i < 0$
then pre $i < 0$
    post $0 \le i \wedge i = -\overleftarrow{i} \wedge j = -\overleftarrow{j}$
else pre $0 \le i$
    post $0 \le i \wedge i = \overleftarrow{i} \wedge j = \overleftarrow{j}$
post $0 \le i \wedge i * j = \overleftarrow{i} * \overleftarrow{j}$

Formally, this argument is also using a rule which permits the use of (stronger pre-conditions or) weaker post-conditions:

$$\frac{PP \Rightarrow P, \ \{P\} S \{RR\}, \ RR \Rightarrow R}{\{PP\} S \{R\}}$$

The post-conditions of the two statements imbedded within the conditional have been chosen to express the intentions of the designer rather than just being copies of *post-MAKEPOS* . In this way—on a problem of greater size—the designer decouples the design of sub-components from their context.

The proof obligation for iteration—as would be expected—is the most interesting. The general form is:

$$\frac{\{P \wedge B\} S \{P \wedge R\}}{\{P\} \text{ while } B \text{ do } S \{P \wedge \neg B \wedge R^*\}}$$

R is required to be well-founded and transitive. The logical expression R is irreflexive, for example:

$$x < \overleftarrow{x}$$

Since the body of the loop might not be executed at all, the state might not be changed by the while loop. Thus the overall post-condition can (only) assume R^* which is the reflexive closure of R—for example:

$$x \le \overleftarrow{x}$$

There is a significant advantage in requiring that R be well-founded (and thus irreflexive) since the above proof obligation then establishes termination. The rest of this rule is easy to understand. The expression P is an invariant which is true after any number (including zero) of iterations of the loop body. This is, in fact, just a special use of a data type invariant. (Notice that such an invariant could fail to be satisfied within the body of the loop.) The falseness of B after the loop follows immediately from the meaning of the loop construct. Returning again to the multiplication example, *POSMULT* might be implemented as in Figure 2. The reader should check carefully how the terms in these logical expressions relate to the proof rule. Notice that rel is well-founded, since i is always positive and cannot be decreased indefinitely (cf. Inv).

```
POSMULT
ext wr i, m: Z, rd j: Z
pre 0 ≤ i
    m := 0
    ;
    pre 0 ≤ i
        while i ≠ 0 do
        Inv 0 ≤ i
            i := i - 1;
            m := m + j;
        rel m = ⃖m + (⃖i − i) * ⃖j ∧ i < ⃖i
    post m = ⃖m + ⃖i * ⃖j
post m = ⃖i * ⃖j
```

Figure 2: Development for Multiplication Example

The implementation in Figure 2 is slow in that it is linear in the value of i. Using the ability of a binary computer to detect the difference between even and odd numbers (by checking the least significant bit) and to multiply or divide by two (by shifting), an algorithm which takes time proportional to the logarithm (base 2) of i is shown in Figure 3. The outer loop of these two algorithms is the same. Notice, however, that the externals clause of *POSMULT* has been modified to permit the necessary assignments to j; the rel clause of the loop has also been changed to cater for the more general case[13]. In both cases the relation is transitive.

It would be reasonable, at this point, to ask how the Inv/rel expressions are discovered. The discovery of proofs from code is not the main objective, and this discussion is avoided. It is, however, possible to observe that the proof step is, in some sense, the inverse of the program design activity. As such it serves as a check in the same way that differentiation of an expression derived by integration is a standard check in the infinitesimal calculus. It is shown in [23] how the proof obligations for programming constructs can be used to stimulate program design steps.

Programs for searching and sorting (cf. [26]) provide many interesting examples for proof construction. In the case of searching, the basic vector involved is not changed and proofs using input/output relations differ little from those which use post-conditions of the final state alone. The utility of the more general post-conditions becomes apparent on sorting examples where the basic vector is changed. Consider the following specification:

```
SORT ()
ext wr l : seq of N

post is-ord(l) ∧ is-perm(l, ⃖l )
```

---

[13]A comparison of these two predicates is actually quite interesting. The earlier one shows directly the remaining work to be done; the latter predicate shows an expression whose value is to be kept constant.

```
POSMULT
ext wr i, j, m: Z
pre 0 ≤ i
    m := 0
    ;
    pre 0 ≤ i
        while i ≠ 0 do
        inv 0 ≤ i
            ext wr i, j: Z
            pre i ≠ 0
                while is-even(i) do
                inv 1 ≤ i
                    i := i/2
                    j := j * 2
                rel i * j = ⃖i * ⃖j ∧ i < ⃖i
            post i * j = ⃖i * ⃖j ∧ i ≤ ⃖i
            ;
            m := m + j;
            i := i - 1
        rel m + i * j = ⃖m + ⃖i * ⃖j ∧ i < ⃖i
    post m = ⃖m + ⃖i * ⃖j
post m = ⃖i * ⃖j
```

Figure 3: Alternative Development for Multiplication Example

The truth-valued function for ordering:

is-ord: seq of **N** → **B**

and that for permutations:

is-perm: seq of **N** × seq of **N** → **B**

are obvious (although it might be interesting to define the latter via its properties rather than directly).

A very simple sorting strategy is to absorb, at each iteration of the loop, the "next" element into its correct position. This design decision can be shown by:

```
SORT
var i: N
i := 1;
while i ≠ n do
inv is-ord(l(1, ..., i))
    i := i + 1
    ;
    BODY(i)
    pre i ∈ dom l
    post l(i + 1, ...) = ⃖l(i + 1, ...) ∧
        ∃j ∈ {1, ..., i} ·
            l(j) = ⃖l(i) ∧ ⃖l(1, ..., i - 1) = del(l(1, ..., i), j)
rel is-perm(l, ⃖l)
post is-ord(l) ∧ is-perm(l, ⃖l)
```

An equally simple (and similarly inefficient) sorting algorithm is one which picks the lowest of the remaining elements and moves it to the next position on each iteration. The additional property is clearly shown in the following invariant:

$SORT$
var $i: \mathbb{N}$
$i := 0;$
while $i \neq n - 1$ do
Inv $is\text{-}ord\big(l(1,\ldots,i)\big) \wedge$
$\qquad \forall m \in \{1,\ldots,i\}, n \in \{i+1,\ldots\} \cdot l(m) \leq l(n)$
$\quad i := i + 1$
$\quad ;$
$\quad BODY(i)$
$\quad$ pre $i \in \text{dom } l$
$\quad$ post $l(1,\ldots,i-1) = \overleftarrow{l}(1,\ldots,i-1) \wedge$
$\qquad \exists j \in \{i,\ldots\} \cdot$
$\qquad\qquad l(i) = \overleftarrow{l}(j) \wedge l(i+1,\ldots) = del(\overleftarrow{l}(i,\ldots),j)$
rel $is\text{-}perm(l, \overleftarrow{l})$
post $is\text{-}ord(l) \wedge is\text{-}perm(l, \overleftarrow{l})$

A comparison of *post-BODY* in the two cases shows the essence of the work to be performed. The development of more efficient algorithms is left as an exercise to the interested reader.

The proof rules shown above cater for post-conditions of two states (input-output relations) but are only slightly more complicated than those in [15] and the examples here have shown that the separation of, for example, invariants and relations can actually aid the design process. (Although the rules in [18] are heavy, it is interesting to see how they can be used as a useful check list in design.)

6 Discussion

This closing section refers to various alternatives to the approach given in the body of the paper. The data refinement rule used above is incomplete in the sense that there are some things that one would like to view as representations which cannot be verified by its use. An alternative rule has been found and is described in [28,13]—a general motivation is included in [23]. The new rule is—in a sense made precise in the source papers—*complete*.

The logic used by this author since [1] caters for partial functions. Essentially, a natural deduction proof style has been provided for the symmetric "three valued" truth-tables of Lukasiewicz found in [25]. A forthcoming paper ([7]) will cover questions of the usability of various approaches to the problem of undefined terms in logical expressions. In [23] reference is made to:

- use of "existential equality" and the problem of its negation;

- use of "strong equality" and separation of definedness proofs;

- approaches due to Abrial, Blikle and Owe.

In this area, it is clear that there is need for further experimentation to establish a convenient proof style.

There are a number of important questions surrounding the relationship between specification and programming languages. In *VDM*'s steps of operation decomposition, for example, it is natural to mix programming language constructs and specifications of operations. Michel Sintzoff (cf. [29]) argues that specification and programming languages are part of a continuum. If this point is conceded, the result should not be to use a programming language as a specification language—the effect is likely to be, at best, a loss of clarity in the expression of partial and non-deterministic operations. More importantly, because of the lack of obvious algebraic properties like commutativity, it is difficult to reason about such texts and this is essential for specifications. Experience with some attempts in this area also suggests that students who are presented with such a language will tend to over use the constructive features.

A more interesting approach is to use parts of a programming language as a subset of a specification language. One way to bring these onto a common semantic footing is to define the programming constructs by translation to predicates. Assignment statements can then be used to overcome the "frame problem" but predicates can be used where more appropriate. Both [16,14], explain approaches which have particularly elegant rules for decomposition. In neither case, however, are they compatible with the satisfaction ordering used here, nor can their notion of specification support as many distinctions concerning non-termination as here. (A very interesting approach is being developed by Jean-Raymond Abrial.) The translation to predicates provides a semantic basis but is likely to lose the link to design decisions. This could result in similar problems to those identified (for example, in [10]) in the use of "Verification Condition Generators". One of the most interesting parts of the work of [16,14] is the development of algebraic properties for the languages involved. This brings together the specification/design/verification and transformational approaches discussed in the introduction to this paper. Perhaps an alternative should be sought in which a restricted subset of predicates can be translated into programs.

Acknowledgements

The author would like to express his thanks to the organizers of the Symposium for their kind invitation to speak in Capri. This paper was typed, under considerable time pressure, by Mrs. Julie Hibbs. The author is also grateful to SERC for financial support.

References

[1] "A Logic Covering Undefinedness in Program Proofs", H. Barringer, J.H. Cheng and C.B. Jones, Acta Informatica, Vol. 21, No. 3, pp251–269, 1984.

[2] "Now You May Compose Temporal Logic Specifications", H. Barringer, R. Kuiper and A. Pnueli, Proceedings of the 16th ACM Symposium on the Theory of Computing, Washington DC, 1984.

[3] "Algorithmic Language and Program Development", F.L. Bauer and H. Wössner, Springer-Verlag, 1982.

[4] "Formal Specification and Software Development", D. Bjørner and C.B. Jones, Prentice-Hall International, 1982.

[5] "A Verification Condition Generator for FORTRAN", R.S. Boyer and J. Strother Moore, pp9–101 in "The Correctness Problem in Computer Science", (eds.) R.S. Boyer and J. Strother Moore, Academic Press, 1981.

[6] "Partial Interpretations of Higher Order Algebraic Types", M. Broy, to be published in the Proceedings of the 1986 Marktoberdorf Summer School, "Logic of Programming and Calculi of Discrete Design", (ed.) M. Broy, Springer-Verlag.

[7] "On the Usability of Logics which Handle Partial Functions", J.H. Cheng and C.B. Jones, forthcoming.

[8] "The Munich Project CIP — Volume 1: The Wide Spectrum Language CIP-L", CIP Language Group, Springer-Verlag, Lecture Notes in Computer Science, Vol. 183, 1985.

[9] "Implementing Mathematics with the Nuprl Proof Development System", R.L. Constable, et al., Prentice-Hall, 1986.

[10] "A Technical Review of Four Verification Systems: Gypsy, Affirm, FDM and Revised Special", D. Craigen, August 1985.

[11] "Fundamentals of Algebraic Specification 1: Equations and Initial Semantics", H. Ehrig and B. Mahr, in "EATCS Monographs on Theoretical Computer Science", Springer-Verlag, 1985.

[12] "Abstract Errors for Abstract Data Types", J.A. Goguen, in: "Formal Descriptions of Programming Concepts", (ed.) E.J. Neuhold, North-Holland, 1978.

[13] "Data Refinement Refined: Resume", J. He, C.A.R. Hoare and J.W. Sanders, ESOP '86, (eds.) B. Robinet and R. Wilhelm, Vol. 213, Springer-Verlag, Lecture Notes in Computer Science, 1986.

[14] "The Logic of Programming", E.C.R. Hehner, Prentice-Hall International, 1984.

[15] "An Axiomatic Basis for Computer Programming", C.A.R. Hoare, CACM Vol.12, No.10, pp576–580, October 1969.

[16] "Laws of Programming: A Tutorial Paper", C.A.R. Hoare, He Jifeng, I.J. Hayes, C.C. Morgan, J.W. Sanders, I.H. Sørensen, J.M. Spivey, B.A. Sufrin and A.W. Roscoe, Oxford University Technical Monograph PRG-45, May 1985.

[17] "Implementation Bias in Constructive Specifications of Abstract Objects", C.B. Jones, 1977.

[18] "Software Development: A Rigorous Approach", C.B. Jones, Prentice-Hall International, 1980.

[19] "Development Methods for Computer Programs including a Notion of Interference", C.B. Jones, Oxford University Technical Monograph, PRG-25, June 1981.

[20] "Specification and Design of (Parallel) Programs", C.B. Jones, Proceedings of IFIP '83, North-Holland Publishing Co., 1983.

[21] "Systematic Program Development", C.B. Jones, in "Mathematics and Computer Science", (eds.) J.W. de Bakker, M. Hazewinkel and J.K. Lenstra, CWI Monographs, North Holland Publishers, pp19–50, 1986.

[22] "Systematic Software Development using VDM", C.B. Jones, Prentice-Hall International, 1986.

[23] "Program Specification and Verification in VDM", C.B. Jones, to be published in the Proceedings of the 1986 Marktoberdorf Summer School, "Logic of Programming and Calculi of Discrete Design", (ed.) M. Broy, Springer-Verlag.

[24] "An Introduction to Proving the Correctness of Programs", S.L. Hantler and J.C. King, ACM Computing Surveys, Vol. 8, No. 3, pp331–353, September 1976.

[25] "Introduction to Metamathematics", S.C. Kleene, North-Holland Publishing Co. Amsterdam, 1967.

[26] "Sorting and Searching", D.E. Knuth, in 'The Art of Computer Programming', Vol. III, Addison-Wesley Publishing Company, 1975.

[27] "Algebra", (Second Edition), S. MacLane and G. Birkoff, Collier Macmillan International, 1979.

[28] "Non-Deterministic Data Types: Models and Implementations", T. Nipkow, Acta Informatica Vol. 22, pp629–661, 1986.

[29] "Expressing Program Design in a Design Calculus", M. Sintzoff, to be published in the Proceedings of the 1986 Marktoberdorf Summer School, "Logic of Programming and Calculi of Discrete Design", (ed.) M. Broy, Springer-Verlag.

Integration of Program Construction and Verification: the PROSPECTRA Methodology

Bernd Krieg-Brückner
Universität Bremen

*A methodology for **PRO**gram development by **SPEC**ification and **TRA**nsformation is described. Formal requirement specifications in Anna are the basis for constructing correct and efficient Ada programs by gradual transformation.*

1 Summary

The PROSPECTRA project aims to provide a rigorous methodology for developing *correct* Ada software and a comprehensive support system. It is a cooperative project between Universität Bremen, Universität Dortmund, Universität Passau, Universität des Saarlandes (all D), University of Strathclyde (GB), SYSECA Logiciel (F), Dansk Datamatik Center (DK), and Standard Electrica S.A. (E), and is sponsored by the Commission of the European Communities under the ESPRIT Programme, ref. #390 (see [1]).

The *methodology* integrates program construction and verification during the development process. User and implementor start with a formal specification, the interface or "contract". This initial specification is then gradually transformed into an optimised machine-oriented executable program. The final version is obtained by stepwise application of transformation rules. These are carried out by the system, with interactive guidance by the implementor, or automatically by compact transformation tools. Transformation rules form the nucleus of an extendible knowledge base.

The strict methodology of Program Development by Transformation (based on the CIP approach, see e.g. [2, 3]) is completely supported by the *system*. Any kind of activity is conceptually and technically regarded as a transformation of a "program" at one of the system layers. This provides for a uniform user interface, reduces system complexity, and allows the construction of system components in a highly generative way.

2 Objectives

Current software developments are characterised by ad-hoc techniques, chronic failure to meet deadlines because of inability to manage complexity, and unreliability of software products. The major objective of the PROSPECTRA project is to provide a technological basis for developing *correct* programs. This is achieved by a methodology that starts from a formal specification and integrates verification into the development process.

The initial *formal requirement specification* is the starting point of the methodology. It is sufficiently rigorous, on a solid formal basis, to allow verification of correctness during the complete development process thereafter. The methodology is deemed to be more realistic than the conventional style of a posteriori verification: the construction process and the verification process are broken down into manageable steps; both are coordinated and integrated into an implementation process by *stepwise transformation* that guarantees a priori correctness with respect to the original specification. Programs need no further debugging; they are correct by construction with respect to the initial specification. Testing is performed as early as possible by *validation* of the formal specification against the informal requirements (e.g. using a prototyping tool).

Complexity is managed by abstraction, modularisation and stepwise transformation. Efficiency considerations and machine-oriented implementation detail come in by conscious design decisions from the implementor when applying pre-conceived transformation rules. A long-term research aim is the incorporation of goal orientation into the development process. In particular, the crucial selection in large libraries of rules has to reflect the reasoning process in the development.

Engineering Discipline for Correct SW: The PROSPECTRA project aims at making software development an engineering discipline. In the development process, ad hoc techniques are replaced by the proposed uniform and coherent methodology, covering the complete development cycle. Programming knowledge and expertise are formalised as transformation rules and methods with the same rigour as engineering calculus and construction methods, on a solid theoretical basis.

Individual transformation rules, compact automated transformation scripts and advanced transformation methods are developed to form the kernel of an extendible knowledge base, the method bank, analogously to a handbook of physics. Transformation rules in the method bank are proved to be correct and thus allow a high degree of confidence. Since the methodology completely controls the system, reliability is significantly improved and higher quality can be expected.

Specification: Formal specification is the foundation of the development to enable the use of formal methods. High-level development of specifications and abstract implementations (a variation of "logic programming") is seen as the central "programming" activity in the future. In particular, the development of methods for the derivation of constructive design specifications from non-constructive requirement specifications is a present focus of research.

The abstract formal (e.g. algebraic) specification of requirements, interfaces and abstract designs (including concurrency) relieves the programmer from unnecessary detail at an early stage. Detail comes in by gradual optimising transformation, but only where necessary for efficiency reasons. Specifications are the basis for adaptations in evolving systems, with possible replay of the implementation from development histories that have been stored automatically.

Programming Language Spectrum: Ada and Anna: Development by transformation receives increased attention world-wide, see [4]. However, it has mostly been applied to research languages. Instantiating the general methodology and the support system to Ada [5] and Anna (its complement for formal specification, see [6]) make it realistic for systems development including concurrency aspects. *PA^{nn}dA*, the PROSPECTRA Anna/Ada subset, covers the complete spectrum of language levels from formal specifications and applicative implementations to imperative and machine-dependent representations. Uniformity of the language enables uniformity of the transformation methodology and its formal basis.

Stepwise transformations synthesise Ada programs such that many detailed language rules necessary to achieve reliability in direct Ada programming are obeyed by construction and need not concern the program developer. In this respect, the PROSPECTRA methodology may make an important contribution to managing the complexity of Ada.

Research Consolidation and Technology Transfer: Research in language design and methodology has traditionally come from Europe; strong expertise in formal methods is concentrated here and has had considerable international influence. Ada will become central for a common European technology base. The PROSPECTRA project aims at contributing to the technology transfer from academia to industry by consolidating converging research in formal methods, specification and non-imperative "logic" programming, stepwise verification, formalised implementation techniques, transformation systems, and human interfaces. A consortium of SYSECA Logiciel and Standard Electrica SA are about to start an ESPRIT demonstrator project of the methodology in an industrial context.

Industry of Software Components: The portability of Ada allows pre-fabrication of software components. This is explicitly supported by the methodology. A component is catalogued on the basis of its interface. Formal specification in Anna gives the semantics as required by the user; the implementation is hidden and may remain a company secret of the producer.

Ada/Anna and the methodology emphasise the *pre-fabrication* of generic, universally (re-)usable, *correct* components that can be instantiated according to need. This will invariably cut down production costs by avoiding duplicate efforts. The production of perhaps small but universally marketable components on a common technology base will not only foster a European market but also assist smaller companies in Europe.

Tool Environment: Emphasis on the development of a comprehensive support system is mandatory to make the methodology realistic. The system can be seen as an integrated set of tools based on a minimal Ada Program Support Environment, e.g. the planned ESPRIT Portable Common Tool Environment (PCTE). Because of the generative nature of system components, adaptation to future languages is comparatively easy. Existing environments only support the conventional activities of edit, compile, execute, debug. Existing transformation systems are mostly experimental and hardly have production quality in user interface, efficient transformation or library support. Conventional verification systems are monolithic and only support a posteriori verification.

The support of correct and efficient transformations is seen as a major advance in programming environment technology. The central concept of system activity is the application of transformations to trees. Generator components are employed to construct transformers for individual transformation rules and to incorporate the hierarchical multi-language approach of PAnndA (PROSPECTRA Anna/Ada), TrafoLa (the language of transformation descriptions), and ControLa (the command language). Generators increase flexibility and avoid duplication of efforts; thus the overall systems complexity is significantly reduced. Choosing Ada/Anna as a standard language, and standard tool interfaces (e.g. PCTE), will ensure portability of the system as well as of the newly developed Ada software. A brief overview of the system can be found in [3]; see also [7-13].

3 The Development Model

Consider a simple model of the major development activities in the life of a program:

Requirements Analysis

- Informal Problem Analysis

- Informal Requirement Specification

Development

⇑ *Validation*

- **Formal** Requirement Specification

⇑ *Verification*

- **Formal** Design Specification

⇑ *Verification*

- **Formal** Construction *by Transformation*

Evolution
- Changes in Requirements ⇒ Re-Development ⇑

The *informal requirements analysis* phase precedes the phases of the *development* proper, at the level of formal specifications and by transformation into and at the level(s) of a conventional programming language such as Ada, see the sections below for discussion.

After the program has been installed at the client, no maintenance in the sense of conventional testing needs to be done; "testing" is performed *before* a program is constructed, at the very early stages by validation of the formal requirement specification against the informal requirements.

The *evolution* of a program system over its lifetime, however, is likely to outweigh the original development economically by an order of magnitude. Changes in the informal requirements lead to re-development, starting with changes in the requirement specification. This requires re-design, possibly by replay of the original development (which has been archived by the system) and adaptation of previous designs or re-consideration of previously discarded design variations.

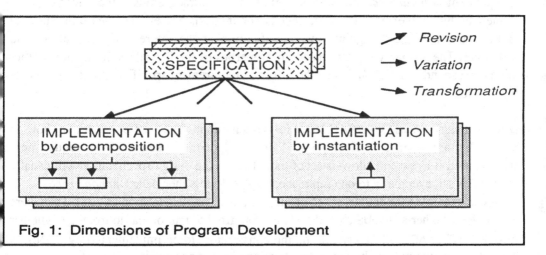

Fig. 1: Dimensions of Program Development

Dimensions of Development: In adaptation of the conventional view of a life-"cycle", one can distinguish several dimensions of program development activities (see fig. 1):

revision	change of a requirement specification to adapt to new requirements
variation	alternate design for the same requirement specification, by
• *decomposition*	("top-down" hierarchy of recursive developments): requirement specification and design of components, or
• *abstraction / instantiation*	(pre-fabrication and use "bottom-up"): generalisation/parameterisation of components to/from stock
transformation	possibly to different language style or language

Specification

A requirement specification defines *what* a program should do, a design specification *how* it does it. The motivations and reasons for design decisions, the *why*'s, should be recorded along with the developments.

Requirement specifications are, in general, non-constructive; there may be no clue for an algorithmic solution of the problem or mapping of abstract to concrete (i.e. predefined) data types. It is essential that the requirement specification should not define more than the *necessary* properties of a program to leave room for design decisions. It is intentionally vague or *loose* in areas where the further specification of detail is irrelevant or impossible. In this sense, loose specification replaces non-determinancy, for example to specify an unreliable transmission medium in a concurrent, distributed situation.

From an economic point of view, overspecification may lead to substantial increase in development costs and in efficiency of execution of the program since easier solutions are not admissible. If the requirement specification is taken as the formal *contract* between client and software developer, then there should perhaps be a new profession of an independent *software notary* who negotiates the contract, advises the client on consequences by answering questions, checks for inconsistencies, resolves unintentional ambiguities, but guards against overspecification in the interest of both, client and developer. The answer of questions about properties of the formal requirement specification correspond to a *validation* of the informal requirement specification using a prototyping tool.

Design specifications specify abstract implementations. They are constructive, both in terms of more basic specifications and in the algorithmic sense. If the requirement specification is loose and allows a set of models, then the design specification will usually restrict this set, eventually to one (up to isomorphism); it is then called *monomorphic*.

As an example take double-ended queues (and their implementation as lists). Then there is a choice whether a DEQUE should be built from front to rear or rear to front, i.e. whether ADD_FRONT or ADD_REAR should be the constructor (to be implemented by CONS on lists). Another example is the loose specification of a priority queue, see fig. 2.

Partial Functions: Fig. 3 shows a monomorphic specification of lists. Just as MIN_ELEM in fig. 2, HEAD and TAIL are *partial functions,* they are only defined if the pre-condition on the parameter holds.

Fig. 4 shows a stack specification. If the **shadowed** pre-condition on PUSH is omitted, STACKS defines unbounded stacks. When we include the pre-condition as shown, PUSH becomes a *partial constructor function;* STACKS then defines bounded stacks. The equations would become partially undefined as well, for some values of S; thus the shadowed premises are included in fig. 4 to insure definedness. A similar example of a bounded data type is INTEGER restricted to an implementation range.

When introducing limitations such as bounds in a methodological step, it is desirable not to have to introduce the definedness premises explicitly; they should be included implicitly by a transformation. In fact it can be argued, that they should not be shown explicitly in the text of the specification at all (see [25]) so that they will not clutter the definition of the "normal cases". They are necessary, however, when reasoning about the definedness of equations during program development and verification, see also section 6 below. The cluttering problem can be solved by different kinds of paraphrasing on the screen, including suppression of the premises as an option (holophrasting).

Subtypes can be used to abbreviate such conditions, possibly leading to a more efficient way of checking for definedness, see fig. 5.

Introduction of Exceptions: Semantically, the introduction of exceptions (see fig. 6) does not alter the specification of the operations within their well-defined domains, that is as long as the pre-conditions hold. The equations remain the same; they are undefined if the pre-conditions do not hold.

```
generic
  type ITEM is private;
  with function "<=" (X, Y: ITEM)
                    return BOOLEAN;
--|for all X,Y: ITEM =>
--|  (X <= X) ∧ ((X <= Y) ∨ (Y <= X)) ∧
--|  ((X <= Y) ∧ (Y <= Z)) -> (X <= Z) ∧
--|  ((X <= Y) ∧ (Y <= X)) -> (X = Y);
package PRIORITY_QUEUES is
  type PRIOQ is private;
  function IS_EMPTY(Q: PRIOQ)
                    return BOOLEAN;
  function ADD(Q: PRIOQ; X: ITEM)
                    return PRIOQ;
  function MIN_ELEM(Q: PRIOQ)
                    return ITEM;
  --| where not IS_EMPTY(Q);
  function REST(Q: PRIOQ)
                    return PRIOQ;
  --| where not IS_EMPTY(Q);
--|axiom for all Q: PRIOQ ; X: ITEM =>
--|  IS_EMPTY(EMPTY) = TRUE,
--|  IS_EMPTY(ADD(Q, X)) = FALSE,
--|  MIN_ELEM(ADD(EMPTY, X)) = X,
--|not IS_EMPTY(Q) ∧ X<=MIN_ELEM(Q)->
--|  MIN_ELEM(ADD(Q, X)) = X,
--|not IS_EMPTY(Q) ∧ X > MIN_ELEM(Q)->
--|  MIN_ELEM(ADD(Q, X)) = MIN_ELEM(Q),
--|REST(ADD(EMPTY, X)) = EMPTY,
--|not IS_EMPTY(Q) ∧ X<= MIN_ELEM(Q)->
--|  REST(ADD(Q, X)) = Q,
--|not IS_EMPTY(Q) ∧ X > MIN_ELEM(Q)->
--|  REST(ADD(Q, X)) = ADD(REST(Q), X);
end PRIORITY_QUEUES;
```

Fig. 2: Loose Specification of Priority Queue

---

f several elements in a sequence of ADD's
are equivalent w. r. t. "<=", an arbitrary one
of these will be chosen by the operation
MIN_ELEM. Similarly, the order in which
elements are added may or may not result in
distinct model representations.

```
package LISTS is
  type LIST is private;
  function NIL return LIST;
  function IS_EMPTY (L: LIST)
                    return BOOLEAN;
  function CONS( X: ITEM; L: LIST)
                    return LIST;
  function HEAD (L: LIST) return ITEM;
  --| where not IS_EMPTY (L);
  function TAIL (L: LIST) return LIST;
  --| where not IS_EMPTY (L);
--|axiom for all X: ITEM; L: LIST =>
--|  HEAD( CONS( X, L )) = X,
--|  TAIL( CONS( X, L )) = L,
--|  IS_EMPTY( NIL ) = TRUE,
--|  IS_EMPTY( CONS( X, L )) = FALSE;
end LISTS;
```

Fig. 3: Unbounded List

In the so-called *initial model* all terms are
considered to be different unless they can be
shown to be equal by equational reasoning.
The PRIOQs

ADD(ADD(EMPTY, X), Y) and
ADD(ADD(EMPTY, Y), X),

have distinct (unsorted) representations and a
search is performed when MIN_ELEM is
applied.

In the *terminal model* all terms are
considered to be equal unless they can be
shown to be different. The two PRIOQs
above have only one, the canonical (sorted)
representation. It is interesting to note that
the operations MIN_ELEM or REST alone do
not allow an *observable* distinction (w.r.t.)
ITEM between the models. Only when we
extend the specification by an auxiliary
operation such as HEAD, we can specify the
different design decisions by requiring

HEAD(ADD(EMPTY,X)) = X,
not IS_EMPTY (Q) →
   HEAD(ADD(Q,X)) = HEAD(Q)
for the initial model, and
   ADD(ADD(Q, X), Y) = ADD(ADD(Q, Y), X),
for the terminal model.

```
package STACKS is
  type STACK is private;
  function EMPTY return STACK;
  function LENGTH (S: STACK)
                      return NATURAL;
  function PUSH (S: STACK; X: ITEM)
                      return STACK;
--| where LENGTH(S) < SIZE;
  function POP (S: STACK) return STACK;
--| where LENGTH(S) > 0;
  function TOP (S: STACK) return ITEM;
--| where LENGTH(S) > 0;
--|axiom for all S: STACK; X: ITEM =>
--|LENGTH(S) < SIZE ->
--|  POP( PUSH( S, X )) = S,
--|LENGTH(S) < SIZE ->
--|  TOP( PUSH( S, X )) = X,
--|  LENGTH( EMPTY ) = 0,
--|LENGTH(S) < SIZE ->
--|  LENGTH( PUSH (S, X)) = LENGTH(S) + 1;
end STACKS;
```

Fig. 4: Unbounded and Bounded Stack

```
package STACKS is
  type STACK is private;
  function EMPTY return STACK;
  function LENGTH (S: STACK)
                      return NATURAL;
  subtype OK_STACK is STACK;
--| where S: OK_STACK =>
--|   LENGTH(S) < SIZE;
  function PUSH (S: OK_STACK;
                   X: ITEM) return STACK;
  function POP (S: STACK)
                   return OK_STACK;
--|  where LENGTH(S) > 0;
  function TOP (S: STACK) return ITEM;
--|  where LENGTH(S) > 0;
--|axiom for all S: OK_STACK; X: ITEM=>
--|  POP( PUSH( S, X )) = S,
--|  TOP( PUSH( S, X )) = X,
--|  LENGTH( EMPTY ) = 0,
--|  LENGTH( PUSH (S, X)) = LENGTH(S) + 1;
end STACKS;
```

Fig. 5: Subtype Abbreviating Condition

```
package STACKS is
  type STACK is private;
  function EMPTY return STACK;
  function LENGTH (S: STACK)
                      return NATURAL;
  OVERFLOW: exception;
  function PUSH (S: STACK; X: ITEM)
                      return STACK;
--|where not (LENGTH(S) < SIZE) =>
--|                raise OVERFLOW;
  function POP (S: STACK) return STACK;
--| where LENGTH(S) > 0;
  function TOP (S: STACK) return ITEM;
--| where LENGTH(S) > 0;
  axioms
end STACKS;
```

Fig. 6: Introduction of Exceptions

```
package STACKS is
  requirement_specification
private -- design specification:
  use LISTS;
  type STACK is new LIST;
--|axiom for all S: STACK; X: ITEM =>
--|  EMPTY = NIL,
--|  PUSH( S, X ) = CONS( X, S ),
--|  POP( S ) = TAIL( S ), TOP( S ) = HEAD( S ),
--|  LENGTH( S ) =
--|    if IS_EMPTY( S ) then 0
--|    else LENGTH( POP( S )) + 1 end if;
end STACKS;
```

Fig. 7: Implementation of (Un)bounded Stack
by Unbounded List

Operationally, however, exceptions allow a delayed test for satisfaction of a pre-condition and therefore an increase in efficiency. For details and transformation of these applicative specification to an imperative level etc. see [15, 16].

Visibility of Equality: In Ada, there is a particularly nasty problem: the implicitly available operation = (when using **private** in the type definition) is not the abstract one as specified in the requirement specification (whether formal or not) but the "identity" on the implementation; this is a clear breach of the information hiding principle. Thus for example two rational numbers would appear to be distinct because they are represented as different pairs 4/2 and 6/3 although they should be equal from an abstract point of view.

The problem is not that one has to define and implement a special operation EQUAL, say, (cf. [5]), but that the operation = on the implementation is still visible, cannot be hidden and may lead to misuses. Overloading of the predefined = is forbidden (this would solve the problem, at least partially). Using **limited private** does not help since = is then hidden but := for assignment or initialisation of variables and constants is hidden as well and cannot be made visible independently from =.

Therefore the operation = is defined to be invisible in PAnndA when using **private**, unless = is explicitly included in the interface (signature) as a virtual operation; the specification must then be monomorphic (that is there is only one model up to isomorphism) and the specified abstract operation = corresponds to *the* = of any implementation. These are technicalities; a revision of Ada is needed in this area, allowing independent visibility, and overloading, of = and :=.

However, = is always visible in specifications and annotations (the Anna part) as the abstract = that is specified in the equations, and is thus available for further specification and verification purposes.

Avoiding Equality: In general, one will specify a special operation EQUAL along with the other operations whenever necessary. However, in many cases equality is not a necessary operation for the user of an abstract type. Complex data structures like files, data bases or symbol tables usually need no test for equality of whole objects, see fig. 3, 4. Thus one can often avoid its inclusion in the interface (as a visible and usable operation in the Ada text) and its explicit specification and implementation, which may be quite involved.

In many cases, a primitive predicate such as IS_EMPTY in fig. 3 is sufficient as an operational test in recursive types.

For the bounded stack, we need an auxiliary operation LENGTH to be able to specify its meaning, see fig. 4. Not only can it be used instead of a function IS_EMPTY as in fig. 3 such that IS_EMPTY(S) becomes LENGTH(S) = 0, but it is essential for the analogous case of a full stack since one needs to reason about the length here. Instead of defining LENGTH as an explicit operation as in fig. 4, on can define LENGTH as an *auxiliary* operation using the symbol --: (see [6]) such that it is only visible in the Anna part of the PAnndA program, eliminating the need for an operational implementation.

5 Development by Transformation

The Transformational Development Model: Each transition from one program version to another can be regarded as a transformation in an abstract sense. It has a more technical meaning here: a transformation is a development step producing a new program version by application of an individual transformation rule, a compact transformation script, or, more generally, a transformation method invoking these. Before we come to the latter two, the basic approach will be described in terms of the transformation rule concept.

A transformation rule is a schema for an atomic development step that has been preconceived and is universally trusted, analogously to a theorem in mathematics. It embodies a grain of expertise that can be transferred to a new development. Its application realises this transfer and formalises the development process.

Transformations preserve correctness and therefore maintain a more formalised relationship to prior versions. Their "classical" application is the construction of optimised implementations by transformation of an initial design that has been proved correct against the formal requirement specification. Further design activity then consists in the selection of an appropriate rule, oriented by development goals, for example machine-oriented optimisation criteria (see Appendix A).

Catalogues of Transformation Rules: Some catalogues of transformation rules have been assembled for various high-level languages. Of particular interest is the structured approach of the CIP group. The program development language CIP-L is formally defined by transformational semantics (see [3, 17, 18]), mapping all constructs in the wide spectrum of the language to a language kernel that is defined by a combination of an Attribute Grammar (see [19]) for the Static Semantics and Predicative Semantics (a variant of Denotational Semantics) for the Dynamic Semantics (see [20]). These basic transformation rules have an axiomatic nature: compact rules for program development can be derived from them in a formal way. A stock of basic PA^{nn}dA transformation rules is the basis for program development in the PROSPECTRA system.

Transformation Rules, Scripts, and Methods: Individual transformation rules are generalised to transformation *scripts*: sets of transformations rules applied together, possibly with local tactics that increase the efficiency. The long term research goal is to develop transformation *methods* that relieve the programmer from considerations about individual rules to concentrate on the goal oriented design activity. A transformation method is thus a set of rules or scripts with a global application strategy.

Language Levels: We can distinguish various language levels at which the program is developed or into which versions are transliterated, corresponding to phases of the development:

- formal requirement specification: loose equational or predicative specifications
- formal design specification: specification of abstract implementation
- applicative implementation: recursive functions
- imperative implementation: variables, procedures, iteration by loops
- flowchart implementation: labels, (conditional) jumps
- machine-oriented implementation: machine operations, registers, storage as array of words

All these language levels are covered by PAnndA (the first two by PAnndA-S), cf.[15, 16, 21, 22].

Appendix A shows a little example carried through these levels down to a flowchart implementation, emphasising the rather "classical" style of Program Development by Transformation as in [2, 3, 17; also 23-26]. It also includes a variation of the design and a revision of the requirement specification.

Current research focuses on the development activities at the specification level, demanding most creativity from the developer. This language level is perhaps the most important programming language of the future (supported by a prototyper, it is a kind of logic programming language). Many developments at lower levels can also be expressed at the specification level, for example recursion removal methods, transforming into tail-recursive functions.

Since the target language is an algorithmic language (Ada) rather than an applicative/functional language with pattern matching (such as Miranda), constructive specifications have to be transformed to algorithmic specifications (see [27]), with selectors on the right-hand side rather than constructors on the left (see LENGTH in fig. 4, 7). Even the (interactive) deduction of constructive design specifications from non-constructive requirement specifications, a kind of *program synthesis,* can be supported by complex transformation strategies and tools. The enhancement of the Knuth-Bendix technique, for example, receives major attention in the PROSPECTRA project (see [28]).

6 Correctness

Integration of Construction and Verification: Not only is the program construction process formalised and structured into individual mechanisable steps, but the verification process is structured as well and becomes more manageable. If transformation rules are correctness- preserving, then only the applicability of each individual rule has to be verified at each step. Thus a major part of the verification, the verification of the correctness of each rule, need not be repeated.

Verification then reduces to verification of the applicability of a rule, and program versions are correct by construction (with respect to the original requirement specification). As the example in Appendix A shows, this proof is much easier than a corresponding proof of the final version.

As an alternative to proving the applicability conditions as they arise, the system can keep track of the verification conditions generated and accumulate them till the (successful) end of the development. This way, no proofs are necessary for "blind alleys", with the danger that the supposedly correct development sequence leading to the final version turns out to be a "blind alley" itself, if the proof fails. But even if we consider all proofs required from the developer (with assistance from the system) together, they are still much less complicated than a monolithic proof of the final version.

Implementation: Certain abstract type (schemata) that correspond to predefined Ada type (constructors), for example **record**, are standard in PA^{nn}dA (see [22]). For example, the usual free term constructions (lists, trees) are available. They can be thought of as being implemented in Ada in a standard way or being turned into an Ada text automatically as an alternative (standard Ada) notation for the package defining the abstract type (see [21] for this notion of "alternative paraphrasing"). We assume that a standard Ada implementation using access types (pointers) and allocators is still considered to be "applicative" at this level of abstraction and that side-effects of allocation will be eliminated during the development process by explicit storage allocation whenever required.

Fig. 3, 4, 7 show the implementation of STACKS using LISTS. The obvious bodies are omitted here; in fact, they can be generated automatically. The operations EMPTY, PUSH, POP and TOP can be identified with NIL, CONS, TAIL and HEAD resp. For LENGTH, an algorithmic recursion equation is now given in the design specification (fig. 7); it could have been generated by transformation from the equations given in fig. 4, see [27]. Since IS_EMPTY is available as an operation on LISTS, it can be used as an operational check to discriminate between the different parts of the domain. All we need to prove for the implementation is the satisfaction of the equation on LENGTH in the requirement specification. This is easily done using for example the first equation, yielding

LENGTH(PUSH(S, X)) = LENGTH(POP(PUSH(S, X))) + 1

and then identifying PUSH(S, X) with some S´.

Notions of Correctness: Various notions of correctness have been distinguished in the literature, in particular in an algebraic framework: total, partial, and robust correctness (see also [20, 29]). It is expected that all three will arise but each has a well-defined place in the methodology.

Partial Correctness: The *revision* of a specification of unbounded stacks to one of bounded stacks as described above (cf. also fig. 4) implies a relationship of *partial correctness* of the latter to the former. A program not using stacks longer than SIZE remains correct under the revision; in general, the pre-condition has to be proved for every call on PUSH to maintain (total) correctness. The generation of such verification conditions upon the revision can be automated by the system.

Total Correctness: Let us now consider the two cases of unbounded and bounded stacks. For unbounded stacks, that is if the pre-condition on PUSH is omitted, the respective operations in STACKS and LISTS can be directly identified as stated. All functions are total; the implementation is *totally correct*. We have considered the special case of a monomorphic requirement specification, that is one that is not loose (there is only one model up to isomorphisms). In general, total correctness preserves the complete set of

Robust Correctness: In practice, we are not so much interested in total correctness; the notion of implementation has to be generalised

(a) to allow a *smaller set of models* for the implementation for a loose requirement specification, and

(b) to allow that operations in the implementation are *more defined* (for example totally defined or raising exceptions) than those in the requirement specification.

The implementation of PRIOQs using unsorted LISTs or, as a variation, sorted LISTs (see fig. 2), is robustly correct according to generalisation (a) above. The implementation of bounded STACKS using unbounded LISTS as shown in fig. 3, 4, 7 is robustly correct according to generalisation (b): the partial constructor operation PUSH is implemented by the total operation CONS, thus it is at least as defined as required.

    As a variation, we could implement a (bounded) stack by a pair of [(constrained) array, index] in the classical way. Here, an unbounded implementation (unbounded array) would have no direct counterpart in Ada but could be used in its abstract form (using a pointer implementation) during the early prototype stages of development.

    Fig. 8 shows the notions of correctness as they relate to our example.

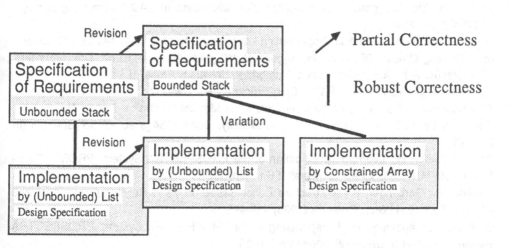

Fig. 8: Notions of Correctness

Correctness of Transformation Rules: In analogy to an algebraic "calculus of data", a transformation rule is an axiom or theorem in a calculus of transformation (cf. also [18]). In fact we can regard the basic transformation rules as axiomatic equations in the algebra of program terms, using algebraic semantics. Alternatively, we can prove the correctness of a basic rule against a given semantics of the (kernel) language.

More complex derived transformation rules actually used in development can then either be proved as equational or inductive theorems or, if the basic rules are loose (or the given semantics is, for example w. r. t. order of evaluation), then transformation rules may introduce design decisions in analogy to design specification (see section 5 above), and are robustly correct. They must, of course, be consistent with the basic rules.

Current research is concerned with such a calculus of transformation rules and their composition to complex development terms, representation of development strategies etc. An issue here is the description of rules that contextually belong together, see [30].

References

[1] Krieg-Brückner, B., Hoffmann, B., Ganzinger, H., Broy, M., Wilhelm, R., Möncke, U., Weisgerber, B., McGettrick, A.D., Campbell, I.G., Winterstein, G.: PROgram Development by SPECification and TRAnsformation. in: CEC, Directorate General XIII (eds.): *ESPRIT '86: Results and Achievements.* Elsevier Sci. Publ. (North-Holland) (1987) 301-311

[2] Bauer, F.L.: Program Development by Stepwise Transformations - The Project CIP. in: Bauer, F. L., Broy, M. (eds.): Program Construction. *LNCS 69,* Springer 1979.

[3] Bauer, F.L., Berghammer, R., Broy, M., Dosch, W., Gnatz, R., Geiselbrechtinger, F., Hangel, E., Hesse, W., Krieg.-Brückner, B., Laut, A., Matzner, T.A., Möller, B., Nickl, F., Partsch, H., Pepper, P., Samelson, K., Wirsing, M., Wössner, H.: *The Munich Project CIP, Vol. 1: The Wide Spectrum Language CIP-L. LNCS 183,* Springer 1985.

[4] Partsch, H., Steinbrüggen, R.: Program Transformation Systems. *ACM Computing Surveys 15* (1983) 199-236.

[5] Reference Manual for the Ada Programming Language. ANSI/MIL.STD 1815A. US Government Printing Office, 1983. Also in: Rogers, M. W. (ed.): *Ada: Language, Compilers and Bibliography.* Ada Companion Series, Cambridge University Press, 1984.

[6] Luckham, D.C., von Henke, F.W., Krieg-Brückner, B., Owe, O.: Anna, a Language for Annotating Ada Programs, Preliminary Reference Manual. Technical Report No. 84-248, Computer Systems Laboratory, Stanford University, June 1984; **also**: *LNCS,* Springer (to appear).

[7] Krieg-Brückner, B.: Informal Specification of the PROSPECTRA System. PROSPECTRA Study Note M.1.1.S1-SN-9.1, Universität Bremen, 1986.

[8] Bertling, H., Ganzinger, H.: A Structure Editor Based on Term Rewriting. in: Proc. 2nd ESPRIT Technical Week, Brussels (1985) 455-466.

[9] Bertling, H., Ganzinger, H.: Paraphrasing in the PROSPECTRA System. PROSPECTRA Report M.1.4-R-1.0, Universität Dortmund, 1986.

[10] Heckmann, R.: An Efficient ELL(1) Parser Generator. *Acta Informatica 23* (1986) 127-148.

[11] Möncke, U., Weisgerber, B., Wilhelm, R.: Generative Support for Transformational Programming. in: Proc. 2nd ESPRIT Technical Week, Brussels (1985) 511-528.

[12] Möncke, U., Pistorius, S., Weisgerber, B.: OPTRAN under UNIX. PROSPECTRA Report S.1.6-R-3.0, Universität des Saarlandes, 1986.

[13] Heckmann, R.: A Proposal for the Sytactic Part of the PROSPECTRA Transformation Language. PROSPECTRA Study Note S.1.6-SN-6.0, Universität des Saarlandes, 1987.

[14] Owe. O.: An Approach to Program Reasoning Based on a First Order Logic for Partial Functions. Research Report No. 89, Institute of Informatics, University of Oslo, 1985.

[15] Krieg-Brückner, B.: Transformation of Interface Specifications. **in**: Kreowski, H.-J. (ed.): *Recent Trends in Data Type Specification.* Informatik Fachberichte 116, Springer 1985, 156-170.

[16] Krieg-Brückner, B.: Systematic Transformation of Interface Specifications: Applicative to Imperative Style, Exceptions. **in**: Partsch, P. (ed.): *Program Specification and Transformation,* Proc. IFIP TC2 Working Conf. (Tölz). Elsevier Sci. Publ. (North-Holland) (to appear).

[17] Bauer, F.L., Wössner, H.: *Algorithmic Language and Program Development.* Springer 1982.

[18] Pepper, P.: A Study of Transformational Semantics. **in**: Bauer, F.L., Broy, M (eds.): Program Construction. *LNCS 69* , Springer (1979) 322-405.

[19] Treff, L.: The Static Semantic Analysis of $PA^{nn}dA$-S; Vol. 1: Rationale, Vol. 2: Attribute Grammar. PROSPECTRA Report S.3.1-R-4.0, Karlsruhe: SYSTEAM KG, 1986.

[20] Broy, M., Nickl, F.: $PA^{nn}dA$-S Semantics. PROSPECTRA Study Note M.2.1.A1-SN-2.1, Universität Passau, 1986.

[21] Krieg-Brückner, B.: $PA^{nn}dA$-S, its Canonical Syntax and Alternative Paraphrasings. PROSPECTRA Study Note M.1.1.S1-SN-7.2, Universität Bremen, 1986.

[22] Kahrs, S.: $PA^{nn}dA$-S Standard Types. PROSPECTRA Study Note M.1.1.S1-SN-11.2, Universität Bremen, 1986.

[23] Broy, M.: Program Construction by Transformations: A Family Tree of Sorting Programs. **in**: Biermann, A. W. et al. (eds.): *Computer Program Synthesis Methodologies.* Proc. NATO Advanced Study Institute, Bonas 1981. NATO Advanced Study Series 95, Dordrecht: Reidel (1983) 1-49.

[24] Partsch, H.: An Exercise in the Transformational Derivation of an Efficient Program by Joint Development of Control and Data Structure. *Science of Computer Programming 3,* 1983, 1-35.

[25] Partsch, H.: Structuring Transformational Developments: a Case Study Based on Early's Recogniser. *Science of Computer Programming 4* (1984) 17-44.

[26] Broy, M., Krieg-Brückner, B.: Derivation of Invariant Assertions During Program Development by Transformation. *ACM TOPLAS 2:3* (1980) 321-337.

[27] Kahrs, S.: From Constructive Specifications to Algorithmic Specifications. PROSPECTRA Study Note M.3.1.S1-SN-1.2, Universität Bremen, 1986.

[28] Ganzinger, H.: Ground Term Confluence in Parametric Conditional Equational Specifi- cations. **in**: Brandenburg, F.J., Vidal-Naquet, G., Wirsing, M.(eds.): Proc. 4th Annual Symp. on Theoretical Aspects of Comp. Sci., Passau '87. *LNCS 247* (1987) 286-298.

[29] Broy, M., Möller, B., Pepper, P., Wirsing, M.: Algebraic Implementations Preserve Program Correctness. *Science of Computer Programming 7* (1986) 35-53.

[30] Qian, Z.: Structured Contextual Rewriting. **in**: Proc. 2nd Int'l Conf. on Rewriting Techniques and Applications, Bordeaux '87. *LNCS* (to appear).

Appendix A

A Little Example of PROgram Development by SPECification and TRAnsformation

A1 Requirement Specification

Several subprograms can be specified in mutual relationship to each other by a (conditional) equational specification as in fig. 9. They are *hierarchically based* on a type defining NATURAL numbers. In other words, the equations do not introduce additional equivalences nor add additional values for NATURAL (cf. [21]).

Both **mod** and "/" are partial functions, the definedness condition has to hold, in this case B > 0. The corresponding definedness premises in the equations (cf. section 4 and fig. 4 above) are considered to be implicit below.

In our example, the two functions are characterised *uniquely* in fig. 9. Had we omitted the clause A **mod** B < B, the results would not be characterised uniquely; however, this would be an under-specification in our particular case (see section 4 above).

The requirement specification is now *frozen* when we proceed to the design.

A2 Design Specification

Constructive Design: The original specification in fig. 9 states desired relationships and computational properties; this is a good style for a requirement specification. However, it is non-constructive, it does not give an indication of how to compute a value.

To develop an implementation for the requirement specification of **mod** or "/", we are now looking for further properties of our operation that characterise a function satisfy

```
package body DIV_MOD is
    function "mod" (A, B: NATURAL)
                          return NATURAL;
    -|  where B > 0;
    function "/" (A, B: NATURAL)
                          return NATURAL;
    -|  where B > 0;
-|axiom for all A, B: NATURAL =>
-|  A = B * (A / B) + A mod B,    A mod B < B;
private
    design_specification
end DIV_MOD;
```

Fig. 9: Requirement Specification

```
package DIV_MOD is
    requirement_specification
private
-|axiom for all A, B: NATURAL =>
-|  A < B -> A mod B = A,
-|  A ≥ B -> A mod B = (A–B) mod B;
-|  A < B -> A / B = 0,
-|  A ≥ B -> A / B = (A–B) / B + 1,
end DIV_MOD;
```

Fig. 10: Design Specification: Integer Division

```
package DIV_MOD is
    requirement_specification
private
    design_specification
-|axiom for all A,B: NATURAL =>
-|  A mod B =    if A < B then A
-|                     else (A–B) mod B end if,
-|  A / B =      if A < B then 0
-|                     else (A–B) / B + 1 end if;
end DIV_MOD;
```

Fig. 11: Algorithmic Specification: Integer Div.

---

ing the specification *constructively* so that we can derive an algorithmic implementation

The properties in fig. 10 are stated as recursion equations; they constitute a design specification for a constructive implementtion. The design corresponds to a simple solution of the problem commonly known as *Integer Division*. Note that we could have come up with a different solution such as *Binary Division* instead, see section A4.

Verification of Correctness: The properties are either additional equations that make a loose specification more specific: then we have to prove consistency w.r.t. the requirement specification, or they are provable as equational or inductive theorems from the requirement specification; this is the case in fig. 9, 10. Such versions can often be derived in a semi-automatic way if the underlying specification (of NATURAL) is contructively specified, by Un/Fold or Knuth-Bendix techniques (see [28]).

The general notion of correctness of implementations is given in the context of algebraic specifications (see section 6 above). Termination is ensured by A – B; the proof is trivial in our case.

A3 Development by Transformation

Algorithmic Design: Now the development by transformation starts. A correct program version is mapped into a correct version if the applicability conditions are satisfied, since the transformation rules have been proved to be correctness-preserving (see section 5 above).

The recursion equations of fig. 10 can be transformed interactively into an algorithmic form, see fig. 11; the set of transformations achieving this is quite involved, see [22].

Applicative Version: Another transformation script achieves the translation of an algorithmic specification into an applicative version in some conventional algorithmic language such as Ada, see fig. 12.

Note that the bodies are generated by transformation and need never be edited. The mapping into a conventional programming language (be it Ada, Pascal, Lisp or such like) is therefore rather discrete and freezes the design specification.

Applicative to Imperative Style: The design activity in each step consists in selecting a transformation rule that achieves a particular purpose, e.g. optimises according to certain criteria. Trafo 1 describes a particularly simple form of recursion-"removal", or rather introduction of the imperative loop construct from an applicative version (see also [17, 26]). Trafo. 1 is only applicable to functions with tail recursion, i.e. functions for which a recursive call is outermost in the expression of the recursive branch in the conditional.

Transformation rules such as trafo 2a, b (or trafo 4 below) can also be expressed at the specification level: the recursion "removal" then corresponds to a mapping from a linearly recursive function (given in recursion equations) to a (system of) tail-recursive functions that can be transformed to loops in an imperative version, see trafo 1a, b. Nevertheless, trafo 2 (and 4) are given to the rather conventional imperative style, incorporating a combination of several transformation steps.

To apply trafo 1a, b to our example, we have to match trafo 1a to fig. 12; it matches for **mod**, not for "/". The syntactic applicability (the matching) is sufficient; in general we may have to prove some semantic applicability condition (the support system will help).

At the same time, substitutions for variables in the program schema on the left hand side of the transformation rule (trafo 1a) are found, e. g. A < B for $B(X)$ (we assume that X stands for an arbitrary number of parameters). The schema variables are then replaced by the program portions on the right-hand side of the schema, trafo 1b; fig. 13 shows the result of the transformation for our example.

```
package body DIV_MOD is
  function "mod" (A, B: NATURAL)
                    return NATURAL is
  begin
   if A < B then
     return A;
   else
     return (A–B) mod B;
   end if;
  end "mod";
  function "/" (A, B: NATURAL)
               return NATURAL is
  begin
   if A < B then
     return 0;
   else
     return (A–B) / B + 1;
   end if;
  end "/";
end DIV_MOD;
```

Fig. 12: Applicative Version: Integer Div.

```
package body DIV_MOD is
  function "mod" (A, B: NATURAL)
                    return NATURAL is
   VA: NATURAL:= A;
  begin
   while VA >= B loop
     VA:= VA–B;
   end loop;
   return VA;
  end "mod";
  function "/" (A, B: NATURAL)
               return NATURAL is
   VA: NATURAL:= A;  VQ: NATURAL:= 0;
  begin
   while VA >= B loop
     VA:= VA–B;    VQ:= VQ+1;
   end loop;
   return VQ;
  end "/";
end DIV_MOD;
```

Fig. 13: Imperative Version: Integer Div.

```
function F (X: S) return R is
begin
  if B(X) then
    return T(X);
  else
    return F(H(X));
  end if;
end F;
```

Trafo 1a: Tail Recursion

```
function F (X: S) return R is
  VX: S:= X;
begin
  while not B(VX) loop
    VX:= H(VX);
  end loop;
  return T(VX);
end F;
```

Trafo 1b: (Tail Recursion) to Iteration

```
function F (X: S) return R is
begin
  if B(X) then
    return T;
  else
    return G( F(H(X)));
  end if;
end F;
```

Trafo 2a: Linear Rec., Constant Termination

```
function F (X: S) return R is
  VX: S:= X;   VY: R:= T;
begin
  while not B(VX) loop
    VX:= H(VX);  VY:= G(VY);
  end loop:
  return VY;
end F;
```

Trafo 2b: (Linear Recursion...) to Iteration

Linear Recursion, Constant Termination: Consider a similar development for "/" instead of **mod** above. As we see in fig. 10-12, "/" is not tail-recursive as **mod** was.

We select another transformation rule for recursion "removal" from our library of transformation rules: trafo. 2a, b. Assuming that its correctness has been proved, we only have to be concerned with its applicability. We find that it is applicable since the result T in case of termination (0 here) is constant in this special case, that is not dependent on the parameters.

Imperative to Flowchart Style: Analogously, we can transform from Imperative to Flowchart Style using trafo. 3a, b, yielding the version of fig. 14.

Development of Transformation Rules: Another observation is that a transformation rule trafo. 1b, c can be derived by applying trafo 3 to trafo 1b, the iteration scheme on the right-hand side of trafo. 1. Furthermore, since correctness-preservation is a transitive property, we can derive a transformation rule trafo. 1a, c from trafo. 1a, b and 1b, c (or 1a, b and 3) by composition of transformation rules.

In this context it is interesting to observe that trafo. 1a, c, mapping directly from tail recursion to flowchart style, can be immediately generalised to several termination cases or several tail-recursive branches without any difficulty, even to a system of mutually tail recursive functions. In contrast, iteration in loops, as in trafo. 1a, b, cannot be generalised easily: multiple termination requires a general condition negating all terminating conditions in the while loop and an extra test

after the loop to distinguish the termination cases; generalisation to mutual tail recursion is impossible.

```
package body DIV_MOD is
  function "mod" (A, B: NATURAL)
                      return NATURAL is
    VA: NATURAL:= A;
  begin
  «REP» if VA < B then goto EL end if;
          VA:= VA–B;
          goto REP;
  «EL»  return VA;
  end "mod";
  function "/" (A, B: NATURAL)
                      return NATURAL is
    VA: NATURAL:= A;   VQ: NATURAL:= 0;
  begin
  «REP» if VA < B then goto EL end if;
          VA:= VA–B;   VQ:= VQ+1;
          goto REP;
  «EL»  return VQ;
  end "/";
end DIV_MOD;
```

Fig. 14: Flowchart Version: Integer Div.

```
function F (X: S) return R is
  VX: S:= X;
begin
«REP» if B(VX) then goto EL; end if;
        VX:= H(VX);
        goto REP;
«EL»  return T(VX);
end F;
```

Trafo 1c: (Tail Recursion) to Flowchart

```
«REP» if not B(VX) then goto EL; end if;
        S(VX);
        goto REP;
«EL»  null;
```

Trafo 3b: (While Loop) to Flowchart

```
while B(VX) loop
  S(VX);
end loop;
```

Trafo 3a: While Loop

```
package DIV_MOD is
  requirement_specification
private
-| axiom for all A, B: NATURAL ⇒
-|  A < B -> A mod B = A,
-|  A ≥ B ∧ A mod (B*2) < B ->
-|    A mod B = A mod (B*2),
-|  A ≥ B ∧ A mod (B*2) ≥ B ->
-|    A mod B = A mod (B*2) - B,
-|  A < B -> A / B = 0,
-|  A ≥ B ∧ A mod (B*2) < B ->
-|    A / B = (A / (B*2)) * 2,
-|  A ≥ B ∧ A mod (B*2) ≥ B ->
-|    A / B = (A / (B*2)) * 2 + 1;
end DIV_MOD;
```

Fig. 15: Design Variation: Binary Division

Transformation Rules and Compilers: Trafo. 3 is the typical example of a transformation rule that would be used in a compiler for the code generation of loop constructs.

The first conclusion we can draw from this observation is that we might as well leave the transformations to and at the flowchart (i.e. the machine-oriented) level to a compiler, as long as we are satisfied with the resulting efficiency. The program developer need only interfere in those cases where more optimisation is called for, say for operations with a high frequency of use. in some cases, interactive optimisation might be called for since the developer has more information about the particular program than an automatic optimiser.

Secondly, code generators and optimisers in conventional compilers can and should be described by (sets of) transformation rules, possibly with the addition of an application strategy, i.e. as *transformation scripts or methods,* see section 5 above. A conventional monolithic optimiser becomes more manageable and flexible if different optimisation techniques and strategies are available as individual transformation scripts. The

```
package DIV_MOD2 is
  type PAIR is private;
  function CREATE (X,Y: NATURAL )
                            return PAIR;
  function QUOTIENT (X: PAIR )
                            return NATURAL;
  function REMAINDER (X: PAIR )
                            return NATURAL;
  function DIV (A, B: NATURAL)
                            return PAIR;
  -|  where B > 0;
-| axiom for all A, B: NATURAL ⇒
-|  QUOTIENT(CREATE(A, B)) = A,
-|  REMAINDER(CREATE((A, B)) = B,
-|  A = B * QUOTIENT(DIV(A, B))
-|        + REMAINDER(DIV(A, B)),
-|  REMAINDER(DIV(A, B)) < B;
private
  design-specification
end DIV_MOD2;
```

Fig. 16: Revision of Requirement Spec.

```
package DIV_MOD is
  requirement-specification
private
  use DIV_MOD2;
-| axiom for all A,B: NATURAL ⇒
-|  DIV(A, B) =    CREATE(A / B, A mod B),
-|  A / B =        QUOTIENT(DIV(A, B)),
-|  A mod B =      REMAINDER(DIV(A, B));
  algorithms
end DIV_MOD;
```

Fig. 17: Design Variation: Embedding

---

transformation support technology in the PROSPECTRA project comes from this area of compiler generation, cf. [11, 12].

So should we map directly to the flow-chart style? Yes, if required. More importantly, we should concentrate our programming efforts on the specification and applicative level, yielding more perspicuous

```
unction F (X: S) return R is
begin
  if B (X ) then
    return T (X ) ;
  else
    return G( X, F(H(X) );
  end if;
end F;
```

such that
```
  function Hinv (X: S) return S
  is an injective inverse function:
    Hinv (H(X)) = X
```

Trafo 4a: Linear Recursion with Inverse

```
package body DIV_MOD2 is
  function DIV (A, B: NATURAL)
                      return PAIR is
  begin
    if A < B then
      return CREATE(0, A);
    else
      if REMAINDER (DIV(A, B*2)) < B then
        return CREATE(
                QUOTIENT (DIV(A, B*2)) *2,
                REMAINDER (DIV(A, B*2)));
      else
        return CREATE(
                QUOTIENT (DIV(A,B*2)) *2+1,
                REMAINDER (DIV(A, B*2))-B);
      end if;
    end if;
  end DIV;
end DIV_MOD2;
```

Fig. 18: Applicative Version: Binary Div2

```
function F (X: S) return R is
  VX: S:= X;    VY: R;
begin
  while not B(VX) loop
    VX:= H(VX) ;
  end loop;
  VY:= T(VX) ;
  while VX / = X loop
    VX:= Hinv(VX) ;
    VY:= G(VY,VX) ;
  end loop;
  return VY;
end F;
```

Trafo 4c: (Linear Recursion) to Iteration

```
package body DIV_MOD2 is
  function DIV (A, B: NATURAL)
                      return PAIR is
    VB: NATURAL:= B;   VP: PAIR;
  begin
    while A >= VB loop
      VB:= VB *2;
    end loop;
    VP:= CREATE(0, A);
    while VB ≠ B loop
      VB:= VB / 2;        -- (2*X)/2=X
      if REMAINDER(VP) < VB then
        VP:= CREATE(  QUOTIENT(VP) *2,
                      REMAINDER(VP));
      else
        VP:= CREATE(  QUOTIENT(VP) *2+1,
                      REMAINDER(VP)-VB);
      end if;
    end loop;
    return CREATE(VQ,VR);
  end DIV;
end DIV_MOD2;
```

Fig. 19: Imperative Version: Binary Div2

programs and, sometimes, even more efficient programs than at the iterative level (when mapped directly to flowchart style).

A4 Variation of the Design

Now let us try a different approach for the design: instead of subtracting B from A to achieve progress in the recursion, let us multiply B by some constant factor 2. We get similar equations for the resulting properties by case analysis as in fig. 10, and after proving their correctness, a recursive version for the Binary Division in fig. 15. We call this a *variation* of the initial design.

Unfortunately, this version of "/" is recursive in "/" and **mod**; so although we could apply trafo. 2a, b, we would have to compute **mod** each time around the loop as well as making the test. So we are stuck with respect to our goal of developing an efficient version.

A5 Revision of the Specification

We should try to combine the two functions into one since they have a very similar algorithmic structure anyway. The "blind alley" of development leads to a *revision* of the original specification to a generalised function DIV with two results: quotient Q and remainder / modulo R, see fig. 16.

The original "/" and **mod** may then be realised as projections QUOTIENT, REMAINDER from the pair of results of DIV as another variation of the design, see fig. 17. The bodies can be generated directly from the algorithmic specifications for "/" and **mod**.

We can now try to relate DIV to our previous designs for **mod** and "/". A standard Un/Fold technique on fig. 17 yields a Binary Division design and the applicative version of fig. 18. Had we chosen the Integer Division design, then our previous development for **mod** and "/" could have been replayed for DIV.

Development of Binary Division: For the Binary Division with two results, we now find that the more general transformation rule for linear recursion, trafo. 4, is applicable: for both X*2 and X*2+1, we can find the inverse X/2; all these operations correspond to simple shift operations on binary numbers. Thus we finally get iterative and flowchart versions for binary division (fig. 19 and 20) corresponding to microcode or even hardware realisations. As a further development in fig. 20, the result variable VP has been represented by two variables VQ, VR, and two independent assignments are performed instead of the collective assignment via CREATE.

Conclusion: Nobody would have wanted to prove the correctness of division in fig. 20 a-posteriori, even for such a little example!

```
package body DIV_MOD2 is
  function DIV (A, B: NATURAL)
                      return PAIR is
    VB: NATURAL:= B;  VQ ,VR: NATURAL;
  begin
  «L1» if A < VB then goto L2; end if;
        VB:= VB *2;
        goto L1;
  «L2» VQ:= 0; VR:= A;
  «L3» if VB = B then goto L6; end if;
        VB:= VB /2;
        if VR >= VB then goto L4; end if;
        VQ:= VQ *2;
        goto L5;
  «L4» VQ:= VQ *2 + 1;
        VR:= VR –VB;
  «L5» goto L3;
  «L6» return CREATE(VQ, VR);
  end DIV;
end DIV_MOD2;
```

Fig. 20: Flowchart Version: Binary Div2

Ada Compiler Validation:
An Example of Software Testing Theory and Practice

John B. Goodenough
Wang Institute of Graduate Studies[1]

Abstract. *The Ada Compiler Validation Capability is a set of tests used to check the conformity of Ada compilers to the Ada Standard. This paper discusses the philosophy and development of these tests from a "lessons learned" perspective, and points out lessons that apply to the testing of other large software projects. Examples of tests and test outputs are given.*

1 Introduction

In September, 1979, SofTech, Inc. started work on the Ada Compiler Validation Capability (ACVC), an effort aimed at developing conformity tests for Ada compilers.[2] In September 1979, only a preliminary specification existed for Ada. A revision was actively underway, resulting in the July 1980 version of the Standard [Ada 80]. We were thus in the position of developing tests for an evolving design. In October 1980, we published a document describing tests that were to be developed for the language [AdaIG 80]. This document was incomplete but provided a basis for initial test development.

In November 1981, revision of the July 1980 specification was started, based on comments and experiences with this version of the language. The ACVC team participated extensively in the revision effort, looking for ambiguities, oversights, and contradictions in the evolving specification. A version of the revised specification was officially issued in July 1982 [Ada 82] and distributed for consideration as an American National Standard. Final revisions were made in January 1983. The specification was accepted as an ANSI/MIL standard February 17, 1983 [Ada 83]. Two implementations of Ada (the NYU Ada/Ed interpreter and the ROLM/Data General compiler) passed the ACVC tests very shortly thereafter (in April and June 1983).

The test suite today (Version 1.8) contains 2857 test files. Since some tests use separately compiled files to check an implementation's treatment of separate compilation, the 2857 files actually represent only about 2400 test programs. The time required to run all the tests ranges from a few hours to several days.

---

[1]On leave of absence from SofTech, Inc.

[2]The effort really should have been called the Ada Compiler *Conformity* Capability since it was focused purely on checking adherence to a Standard, but ACVC is the established term.

The test suite is not yet complete. We have identified approximately 1400 additional tests that need to be written just to cover all aspects of the language adequately. Despite its incompleteness, the tests provide a searching coverage of much of the language.

The purpose of this paper is to discuss some of the "lessons learned" in developing the ACVC tests, and in particular, to point out lessons that apply to the testing of other large software projects.

2 Compiler Testing vs. Testing Other Kinds of Software

Developing tests to check a compiler's conformance to its specification is different in many ways from developing tests for other kinds of software. But there are many important similarities that make the lessons learned in developing ACVC tests applicable to other efforts. The main similarities are:

- *the complexity of the software being tested:* there are many language features to be tested so the number of tests to be developed is very large. Moreover, the features interact in potentially unexpected ways, so tests must be written to check that combinations of features are handled correctly.

- *long life:* compilers have a long life. More effort can be spent in developing tests because the cost can be amortized over a long period.

- *independent test development:* the ACVC test developers were completely independent of those who were developing compilers. Increasingly often in industry today, acceptance tests for software are developed by an independent quality assurance (QA) group that is responsible for deciding whether a software product meets its specifications and is ready for release.

- *implementation on different hardware:* different implementation teams develop Ada compilers for different computers. The tests must be designed to accomodate allowable implementation differences, especially differences that derive from underlying hardware differences. Software developed for execution in a variety of hardware environments has similar testing requirements.

The lessons learned in the ACVC development effort apply to the development of tests for any software having the above characteristics.

There are important differences between the ACVC effort and a normal internal QA testing effort. Some of these differences, and their importance are:

- *stability of the specification:* The Ada specification was, in principle, unchangeable. In contrast, when implementation problems are discovered during a normal internal development or when developing software under contract, the specification can often be adjusted to alleviate the problems. For example, implementers might discover that certain requirements seem to be overly expensive to implement when compared with potential benefits to the customer. Modifications to the specification that reduce implementation costs can then be considered. Similarly, after substantial implementation effort has been spent, the QA group might discover that the implementers have misunderstood the specification. Once again, it is usually possible to salvage some of the implementation effort by modifying the specification. In short, working to a truly frozen specification is unusual. (In practice, the tests have evolved in response to some of these pressures. More specifically, the Ada Language Maintenance Committee has sometimes recommended interpretations of the Standard that reflect implementation difficulties, implementer interpretations, or difficulties experienced by users.[3] Some interpretations have been motivated by the need to have efficient object code.[4] In at least one case, an ambiguity in the Standard was resolved by determining how most implementations had interpreted the Standard.)

- *black box testing only:* The ACVC tests are black box tests because vendors cannot be expected to deliver source code for testing, and even if they did, the process of instrumenting it would be impractical. In contrast, internal testing teams often have access to source code. If they wish, they can instrument the code being tested to evaluate the thoroughness of the testing process, augmenting tests based on an assessment of what parts of the code have been exercised. This option was not available to us and so did not affect the testing approach. (On the other hand, testing instrumented code is so expensive that it is not usually done in any case.)

- *emphasis on conformity testing:* an internal QA group has a broader goal than merely assessing whether software satisfies its specifications. The QA group also exists to ensure the software meets user needs. Evaluating usability is more difficult than evaluating conformity. It is an important limitation that ACVC tests generally did not attempt to evaluate compiler usability.

- *requirement to pass all tests:* the sponsor of the ACVC effort, the U.S. Department of Defense (DoD), specified that all tests had to be passed before a compiler would be considered acceptable for use within DoD.

---

[3]The Ada Language Maintenance Committee (LMC) was established formally by the group responsible for developing the International Standard for Ada. (This group operates under the auspices of the International Organization for Standardization. Its official designation is ISO/TC97/SC22/WG9, known informally as WG9.) The LMC recommends interpretations of the proposed International Standard to WG9. If WG9 accepts the recommendations, they become official for the International Standard. There is a corresponding committee (known as the Ada Language Maintenance Panel, or LMP) that recommends interpretations of the ANSI-MIL Standard. The LMC and LMP are, in practice, identical committees, but they have a different official status. The LMP has only recently been established officially, and operates under the auspices of the Ada Board, a Federal Advisory Committee. The LMP's recommendations are first considered by the Ada Board, and if approved, are then considered by the Director of the Ada Joint Program Office, who is the only person authorized to issue official interpretations of the ANSI-MIL Standard.

[4]For example, interpretation AI-00387 allows implementations to raise CONSTRAINT_ERROR when the Standard specifies that NUMERIC_ERROR should be raised. This interpretation was developed after several implementers noted adverse efficiency consequences of the original rules and after it was noted that programmers could seldom depend on which of these two exceptions would be raised anyway.

This criterion is stronger than that used in many organizations. A product can often be released even with known errors if it is felt the errors will not seriously affect potential users.

3 Early Focus on Testing

DoD's goal in sponsoring the Ada effort was to produce a single programming language that would meet DoD's programming needs [Carlson 81]. DoD had had experience with attempting to standardize languages for military use and had found that it was not enough simply to promulgate a standard. Implementations inevitably diverged from a mandated standard if there was no mechanism for detecting deviations. Some of the reasons for deviations were:

- misinterpretation of the standard;
- a decision on the part of an organization paying for compiler development that certain expensive features of the language did not need to be implemented given the anticipated needs of the application; such decisions led to subset implementations;
- perceived failings in the language that were corrected unilaterally by implementers (leading to superset implementations or deviant interpretations of language constructs).

The effect of project pressures to produce specialized versions of compilers made the need for compiler conformity tests obvious. Otherwise, DoD's long range, project-independent goals would not be met. It was also obvious that an effective enforcement mechanism needed to be in place before the first implementation was delivered.

To ensure early availability of conformity tests, DoD started test development early, in parallel with refinement of the language design. It is not often the case that an independent test team is invited to participate in reviews of the specification that is to be tested. But such a role is certainly important, both in helping to ensure the correctness of the tests (testers have to interpret specifications correctly, just as implementers do!) and in helping to improve the consistency and precision (testability) of the specification.

During the refinement of Ada's design, the ACVC test team served an important function by identifying inconsistencies and imprecisions in the specification drafts. The design team and other reviewers were heavily occupied with evaluating the design for completeness (the intended functionality was specified) and utility (the specified functionality would fit user needs). From November 1981 to January 1983, over 5000 written comments on drafts of the design were sent to the language design team. 25% of these comments were submitted by the ACVC team.

It is clear that involvement of the test team early in the specification process was very productive in reducing the number of errors that exist in today's Standard. Designers are heavily concerned with ensuring that essential functionality is specified. Checking the details is difficult, in part because the intent behind any wording in a specification is often much clearer to the person who writes the specification than to the reader. Having an arms-length review by people who were primarily concerned with specification precision (rather than with functionality) was important; unless people are specifically assigned this role, not enough attention will be paid to precision. In addition, it is difficult for a designer to detect inconsistencies that are introduced as a design evolves. Having specific people focus on detecting inconsistencies (rather than worrying about their significance) is very helpful in producing an internally consistent specification.

4 Conformity Testing vs. Quality Assessment

The ACVC tests are conformity tests, i.e., they check whether an implementation conforms to the requirements of the Standard. Conformity to the Standard in no way implies that a compiler is useful. It may compile slowly; the size of programs it can process may be very limited; its maintainability may be poor. The fact that the ACVC tests were not aimed at evaluating the *quality* of a compiler was mentioned in our original proposal to the DoD, and was emphasized again in [Goodenough 81].

DoD's emphasis on conformity testing was understandable, but has proven to be short-sighted. Tests for evaluating quality are equally important. DoD somehow assumed it would be sufficient just to have conformity tests. How did this happen?

Although I did not participate in any decisions or conversations about whether DoD should support a compiler quality assessment effort in parallel with the ACVC effort, it is easy to see why DoD gave little emphasis to the quality assessment problem (and easy to see how other organizations could make the same mistake). In DoD's case, the *primary* objective was to create a standard language for DoD use. Experience clearly showed there would be no DoD common language unless implementations were checked for conformity (and DoD required the use of conforming compilers). So the emphasis on conformity testing is entirely understandable. There was no other way to ensure the use of a truly common language within DoD.

In addition, there was growing experience with the COBOL and FORTRAN test suites [Oliver 79, Hoyt 77], and the Pascal test suite was under development [Wichmann 83]. All these test suites were aimed at evaluating conformity to a standard.[5] Why should anything more be needed for Ada? As it turned out, DoD's emphasis on passing the tests (see the next Section) and the nature of the tests themselves meant that as some implementers approached their initial deadlines for compiler validation, more effort had to be spent on passing the tests than on improving performance or on removing bugs that were not detected by the tests. When software development is test-driven, it is important

---

[5]The Pascal test suite is partially an exception. It has some tests that report aspects of an implementation's quality, e.g., there are tests that determine what floating point precision is supported.

to be sure the tests correctly reflect all important user needs. It was certainly not anticipated that initial deliveries of Ada compilers would be so driven by the validation tests.

In retrospect, it is easy to see that DoD should have started a compiler quality evaluation effort in parallel with the conformity testing effort. Members of a quality assessment team would have focused on identifying properties of implementations that would make them more usable for typical DoD projects. Criteria (and evaluation tests) could have been constructed for checking:

- object code quality (size/space), including optimization options supported by an implementation

- compilation speed (which is important when there are millions of lines of code to be compiled, and recompiled)

- helpfulness of compile-time error messages

- size of compilable programs (e.g., more than 400 lines per compilation unit?)

- support for implementation options
    - single/double precision?
    - maximum number of tasks that can be created
    - number of levels of priority
    - support for interrupts and representation clauses
    - storage allocation strategy
        - no storage recovery
        - garbage collection
        - programmer-controlled deallocation

The usability of a conforming compiler for a particular project would depend on how well it performed against these and other criteria.

It is important to give usability as much attention as conformity because:

- usability is not usually addressed explicitly in specifications, particularly in language specifications. (It would indeed be considered strange to see requirements specifying, say, the minimum size of programs that could be compiled.)

- usability requirements differ depending on user requirements (e.g., the target computer architecture and the proposed use of the compiler). In particular, there are no absolute measures of quality. Each measure of quality must be weighed by a potential user to decide whether a particular conforming implementation has made appropriate tradeoffs.

- defining usability evaluation criteria is hard. It will not be done unless specific attention is given to it.

There would have been little overlap between usability evaluation and conformity evaluation. A team aimed at evaluating conformity assesses a design's consistency and precision; these properties are orthogonal to usability.

5 Importance of Passing the Tests

Because of DoD's past experience with non-standard implementations of standard languages, a firm policy was defined for Ada — an implementation must pass all applicable correct tests, or be considered nonconforming, *and unusable on DoD projects*. Failure to conform now had economic consequences. The ACVC tests now became the stamp of approval needed to tap the DoD marketplace.

This measure of conformity was much more severe than that used for other languages. In the case of COBOL, FORTRAN, and Pascal, if an implementation fails some tests, these are noted and potential users are allowed to decide for themselves how important the failures are.[6] Some failures might not have any practical consequences for a projected use of the language.

DoD's emphasis on demonstrating conformity by passing the ACVC tests had many important and sometimes unexpected effects:

- DoD's control of the testing process meant potential Ada users looked at a compiler's validation status as a prerequisite to purchase or use. From a vendor's viewpoint, it appeared that only DoD-validated compilers would be saleable, so ...

- implementers took the specification seriously, even the parts that were hard to implement or that made compilers less efficient.

- as the only method for evaluating compilers, the conformity tests were (initially) assumed to include warranties of fitness for use. This impression was fostered by calling the conformity tests "validation" tests.

- because of the importance of passing the conformity tests, implementers trying to make their first delivery focused on passing the tests rather than on removing bugs that weren't detected by the tests or on improving compiler usability. This delayed the release of partial Ada compilers, and delayed experimentation with the language. It also meant that initial deliveries of Ada compilers tended to perform more poorly than initial deliveries for other languages (since the initial deliveries for other languages didn't have to demonstrate strong conformity to a standard).

- the economic importance of passing the tests meant that implementers tended to assume the tests were correct when dealing with marginal interpretations of the language. This had the positive effect of making implementations agree on an interpretation, but was negative when an implementer later challenged a test successfully, causing the test (and future versions of compilers) to be changed.

The impact of the tests on the initial usability of compilers was perhaps made more

---

[6]In practice, implementers strive to pass all the tests. In experience with the validation of Pascal compilers, only one certificate has been issued for a system that did not pass all the tests [private communication, Brian Wichmann].

severe because of the philosophy of test development, which is discussed in the next section.

6 Test Structure and Approach

The development of the ACVC tests has been guided by a document called *The Ada Compiler Validation Implementers' Guide* (AIG) [AdaIG 80, AdaIG 86]. This document analyzes and summarizes the rules given in the Standard and specifies objectives for individual ACVC tests.[7] The "implementers" mentioned in the title are thus primarily the implementers of tests, not the implementers of compilers. This means the document does not discuss implementation techniques that are particularly suitable for Ada compilers but instead identifies the sometimes subtle consequences of the Standard's rules so appropriate tests can be written. Of course, these consequences are also of interest to compiler developers.

When the ACVC contract started, it was envisioned that the AIG would be quickly completed and made available to warn implementers of potential implementation subtleties while they still had time to do something about them. Unfortunately, the amount of time required for reviewing the evolving language and the difficulty of exploring the consequences of language rules that were changing caused considerable delay in completing the AIG. Although an incomplete version of the AIG was published in 1980 [AdaIG 80] (for the July 1980 version of the language), a complete version of the AIG for the 1983 Standard will not appear until December 1986 [AdaIG 86], although portions of this document have been available since 1983.

The delayed completion of the AIG did not have much effect on the pace of test development; it was almost always possible to stay well ahead of the test implementers. Unfortunately, there was one time when a lack of test objectives threatened to leave test developers idle. When this problem was recognized, test objectives were written quickly without the usual careful analysis of what should be tested. As would be expected, tests developed for these hastily written objectives have proven to be much weaker than those created for the rest of the language — a number of important potential compiler deficiencies were overlooked. Moreover, when the full analysis was done later, it proved difficult to integrate the existing tests with test objectives that were then seen to be appropriate. As usual, hasty design (whether for tests or for software systems) is never cost-effective in the long run.

---

[7]A sample section from the AIG is given in Appendix I.

6.1 Organization of Tests

As described in [Goodenough 81], the tests followed the Pascal test suite approach by keying each test to a section of the Standard. When a test was intended to check interactions between features described in different sections of the Standard, test objectives were written in both sections, but only one test was written.

Tests were divided into categories (examples from each category are given in the Appendices):

- non-executable
    - errors to be detected at compile time (Class B tests)
    - errors to be detected at link time (Class L tests)
- executable
    - capacity tests (Class D tests) — the test may exceed the capacity of an implementation, and thus be inapplicable[8]
    - other executable tests (Class A and Class C tests; the difference between these classes is no longer relevant)
- tests whose pass/fail criteria were special or that required special processing of some kind (Class E tests)

The class of each test was indicated in its name so it would be clear whether a test was expected to compile and execute. Executable tests were self checking and printed PASS/FAIL messages in a standard format that made "failed" messages stand out (see Appendix IX). Examples of each class of test are given in the Appendices.

The definition of class E tests gradually evolved as it was discovered that some tests required special interpretation. For example, it was discovered that some implementations detected certain errors at link time while others issued compile-time error messages. Placing these tests in a special class ensured they would receive appropriate attention. Also, the Standard sometimes allows considerable variation in implementation behavior. Special analysis of test results is sometimes needed to see if the implementation has stayed within the allowed range of variation. (An example of a class E test is given in Appendix VI. The test can be rejected at compile-time by an implementation, or an implementation can accept and execute the test. If an implementation rejects the test, the output must be examined to see if the rejection is for an allowed reason (see Appendix VIII).)

---

[8]The distinction between a capacity test and a conformity test is not always obvious. For example, the Standard gives no lower or upper limit on nesting levels. Should a compiler be considered nonconforming if it rejects programs with loops nested to a depth greater than 7? 16? 1,000? The Standard specifies no minimum size for compilation units that must be accepted. Can a conforming compiler refuse to process a compilation unit that has more than 400 statements? more than 1,000 declarations? The decision we made was to test some of these limits and report when certain limits were encountered. Thus an implementation could fail a capacity test and still be considered validated, but the implementation's failure would be reported in the Validation Summary Report.

6.2 Test Selection and Design

The systematic development of compiler conformity tests poses great difficulties. It is clearly impractical to try to test all combinations of language features. Judgement must be used to select those combinations that are considered most important to test.

As stated in [Goodenough 81], tests were constructed with the goal of demonstrating the absence of certain kinds of implementation errors. In particular, when designing test objectives and tests, we gave attention to:

- parts of the language that were hard to implement, either because considerable effort would be required or because the language rules were complex. If there was a penalty associated with implementing the language correctly, we wanted to be sure every implementer paid it. An implementer who analyzes the language carefully and goes to considerable trouble to handle difficult cases correctly should be rewarded by being able to pass a test that a less careful competitor might initially fail. Writing such tests encouraged implementers to take the specification seriously.

- potential implementation oversights. Some interactions between rules might be overlooked even by a careful implementer. Tests were designed and written to detect such oversights. Often these oversights dealt with "boundary" conditions of various kinds; from an implementer's viewpoint, these test cases often looked pathological, i.e., correct with respect to the Standard, but useless to programmers and hard to implement right because they introduced special cases into the compiler.

- denial of useful features. We wanted to be sure that features likely to be used by programmers worked correctly. Tests written from this viewpoint did not focus necessarily on special cases but instead just verified that some feature seemed to work correctly.

- extensions to the language. It is not, of course, possible to check systematically for extensions to a language, but tests can be written to ensure that implementers enforce rules they might consider distasteful or unintended on the part of the language designers. (For example, to help preclude extensions, a test was written to check that only the specified identifiers were considered reserved words. This test checked that none of the reserved or key words from widely used languages or from DoD's languages were treated as reserved words unless they were already reserved by the Standard (see Appendix III).[9] If nothing else, this test ensured that many attractive identifiers could not be used as reserved words if an implementation wanted to support an extension to the language.)

The effect of this test development approach was not considered entirely positive by

---

[9]The languages considered were: COBOL, FORTRAN, PL/I, ALGOL 60, ALGOL-68, Pascal, JOVIAL (J3, J3B, J73, and J73/I), TACPOL, SPL/I, and CMS-2.

some implementers because the decision to check boundary conditions and interactions between rules led to a greater number of unusual (or "pathological") tests than might otherwise have been the case. (A "pathological" test checks a condition that might be considered unlikely to occur in practice, even for programs that are generated by macro processors or other forms of code generators.)   Some people associated with implementations have complained (usually privately) that such tests delayed compiler deliveries for as long as a year while errors discovered by the tests were corrected, but because the tests were "pathological," the corrections (in their opinion) did not make the compilers more useful to customers.   Certainly having to spend time getting peculiar cases right meant implementers could devote less time to making compilers run more efficiently for those capabilities of interest to their customers. Thus, compiler usability was sometimes traded off in favor of passing the tests.

In fact, it was somewhat surprising to find that interactions between "orthogonal" rules in a specification can produce cases that are difficult to implement and that seem unlikely to be useful in practice. For example, objects of any type, including task types, can be declared locally in functions, and functions can return values of any type. A function containing a locally declared task cannot be exited until the local task has completed its execution (to ensure that any non-local storage accessed by the task is not deallocated until the task no longer needs it). The combination of these rules allows a function to return a task that has been declared locally within it, but such a task cannot be used usefully for any purpose, since it is always terminated. For example:

```
package P is
    task type Tsk is
        ...
    end Tsk;
end P;

with P;
function F return P.Tsk is
    Local_Task : P.Tsk;
begin
    ...
    return Local_Task;
end F;

...F'Callable      -- always false
...F'Terminated    -- always true
```

Because of the possibility of using the function call in an attribute, an implementation cannot destroy the task control block created for F.Local_Task. A somewhat more complex implementation is needed just to handle the truly unlikely possibility of checking the callable status of a task returned by a function.

When such interactions are noticed, the tester might decide against writing a test case, allowing some implementers to provide a simpler implementation that doesn't support these unusual cases correctly. It may be that the nonconformity will never be noticed in practice. This approach would have the effect of making the specification appear simpler than it really is because writing explicit rules that outlaw the hard-to-implement cases would introduce special cases into the specification, making it more complex.

Instead of focusing on potential implementation errors, an alternative test development approach would have been to focus on anticipating user needs, writing tests to check that common user errors would be detected and that all important functionality was present. After developing a full set of tests according to this criterion, one could then add tests to check implementation failures as they were reported in practice. With this approach, compilers might not implement certain borderline aspects of the language correctly, but this wouldn't matter if no user of the language wrote borderline programs. Tests written under this philosophy would not search for particular kinds of implementation errors, but would be written simply to obtain evidence that expected functionality was present. Of course, just as it is impossible to write tests that detect all possible implementation errors, it is impossible to write tests that properly anticipate user needs, particularly for a new programming language where usage patterns have yet to be established. With either test approach, a compiler that passes the tests will still contain errors, but perhaps with the alternate approach, the errors that are present would have little effect on the kinds of programs that are actually written.

In retrospect, the job of designing and implementing tests was easier because we did not give any consideration to the potential importance of possible implementation errors, but this approach led to a test suite in which many tests check conditions that are probably not of much interest to actual users. An alternate testing philosophy has been suggested by Mills: "Testing is not a development activity for discovering product defects but an independent assessment of software execution in representative operating environments [with the goal of making statistical inferences about the expected reliability of the software in actual operation]" [Currit 86]. Mills advocates using a sample of expected user inputs as the test suite. Of course, obtaining a statistically acceptable sample is not really possible when the software being tested is a compiler for a new programming language, and even if such a sample could be constructed, it would have to consist of many more tests than are anticipated for the ACVC. But the essence of the Mills' philosophy is to select tests that reflect anticipated user needs rather than possible implementation errors. This approach was not even considered in designing and implementing the ACVC tests, but it could have had some advantages.

6.3 Test Size

In my initial plan for organizing the tests, the intent was to have many small tests. This intent has generally been achieved. There are about 92,000 declarations and statements in Version 1.8 of the test suite (computed by counting occurrences of semicolons). These are distributed among 2857 individual test files, so we have, on average, 32 declarations and statements per test, indicating that the average test is rather simple. Counting physical lines instead of semicolons, the 2857 test files contain about 222,000 lines; the average file contains 77 lines. The distribution of test size is quite skewed: 25% of the tests have fewer than 10 declarations/statements and are less than 30 lines long. 50% of the tests have fewer than 58 lines (22 statements/declarations). The longest test has 811 lines, with 766 statements/declarations. The distribution of declarations/statements by test class is:

Class	# tests	semis	mean	median
Class A	71	2,705	38	30
Class B	1,026	24,079	23	12
Class C	1,519	61,946	41	31
Class D	50	2,277	46	19
Class E	24	393	16	16
Class L	167	786	5	4
Totals	2,857	92,186	32	22

The distribution by number of lines is:

Class	# tests	lines	mean	median
Class A	71	5,469	77	72
Class B	1,026	57,675	56	36
Class C	1,519	149,624	99	77
Class D	50	4,500	90	56
Class E	24	1,117	47	52
Class L	167	3,187	19	14
Totals	2,857	221,572	77	58

The use of fairly small tests has had one unexpected benefit — when a test is discovered to be in error and withdrawn from the test suite, there is only a small overall impact on the thoroughness of the tests, since only a few concepts are checked in any given test. The use of small tests also helped implementers, who could identify and work on more than one error at a time.

6.4 Implementing the Tests

Identifying test objectives is one issue; another is deciding in what order to write the tests. Since the ACVC goal was to have a test suite available as soon as possible after final approval of a Standard, and since the Standard was being revised in important ways after July 1982 and before final approval in February, 1983, it was difficult to meet this goal. In particular, it was necessary to write tests as early as 1980, before revision of [Ada 80] had even begun.

In practice, it turned out that few tests written between 1980 and 1983 had to be discarded, mostly because we chose to implement tests for that part of the language that seemed least likely to change. Since we were intimately involved with the design revision, it was easy to decide which areas of the language were stable. Of course, this test development approach meant that some important areas of the language were untested in early releases of the suite because the final resolution of certain design issues was too uncertain.

Other criteria were also used in deciding which tests to implement. Although a conscious decision was made to defer the implementation of some tests because the issues being tested were unlikely to be of much practical interest, and although unimplemented test objectives are now prioritized in terms of the importance of ruling out certain errors, in practice, the selection of test objectives to implement was determined by the availability of test objectives, the ease of writing the test, the experience level of test implementers, and the efficiency of having a test developer focus attention on a particular area of the language (implementing all the tests relevant to that area regardless of the relative importance of the test objectives).

It is, of course, not enough just to write tests. They must be checked just like any other kind of software. Although every test is given an independent review, the independent review would not have been sufficient to ensure test correctness. Fortunately, we had access to the NYU Ada/ED Interpreter, and this implementation was (and is) used during test production to check that the tests themselves seem to be correct.[10] The use of Ada/Ed was mutually beneficial: we discovered errors in their implementation, and Ada/Ed discovered errors in the tests. Of course, sometimes Ada/Ed failed to detect an incorrect test. Such tests were revealed when the tests were used by other implementers.

Not all tests in the suite were written by the test team. Implementers and users alike have contributed tests. Implementers have sometimes contributed tests after observing weaknesses in the validation suite. Such contributions tend to be in areas that were difficult to implement correctly. After taking the time and effort to do it right and after writing tests themselves to check their implementation, these tests have been submitted to the ACVC team for consideration, and have often been added to the suite. Sometimes test objectives have been added as well. Users have also contributed tests after observing weaknesses in the test suite. Both sources of tests have been valuable in improving the ability of the suite to detect implementation errors.

The test suite is updated continually and released periodically. Updates are needed to correct errors in tests. In addition, new tests are added, and sometimes existing tests are strengthened. Sometimes tests have to be changed because of interpretations recommended by the Language Maintenance Committee/Panel.

---

[10]Other compilers are now also used. Each released test is checked by at least two compilers. More are used if no compiler seems able to pass a test.

The official version of the test suite is currently in effect for a 12-month period (although prior to June 1986, it was in effect for only six months and new versions were released every six months). A version of the test suite is released for public review six months before it becomes the official version. For example, version 1.9 of the suite will be released for review on December 1, 1986. It will become the official version on June 1, 1987 and will be in effect (i.e., usable for validations) until May 30, 1988. Version 1.10 will be released for review December 1, 1987, etc.

7 The Testing Process

Because an implementation had to pass all applicable tests to be considered conforming, and because having a conforming compiler was economically important to vendors, vendors were understandably nervous about the compiler testing process. It was decided early-on that vendors would be given copies of the tests in advance and a formal check of conformity would be scheduled only when the vendor had passed all the tests internally. Formal testing (under supervision of an independent testing agency) was expected to be *pro forma*. This expectation proved faulty almost immediately when it was discovered that vendors sometimes overlooked test failures, either by oversight or because the criteria for passing some tests were misunderstood. In addition, it was not always immediately clear whether an implementer was taking advantage of a permitted implementation option or not.[11] For example, consider the follow fragment of a test:

```
case X is
    when -5_000 => ...
    when       1 => ...
    when 10_000 => ...
end case;
```

One vendor implemented all case statements with jump tables. When compiling case statements like the one given above, this implementation ran out of space to store its jump table. The implementer argued that this was an acceptable "capacity" limitation, but his argument was not accepted; to pass the tests, he had to provide an alternate case statement implementation that was more space-efficient for such examples. The same implementation raised USE_ERROR when any attempt was made to create a sequential or direct file. In effect, the implementation did not support sequential and direct I/O. It was decided that this was a permitted implementation option, although a more strict rule was later imposed — sequential and direct I/O must be supported unless such support was essentially impossible, e.g., because there are no disk or tape drives.

---

[11]An implementer is always allowed to challenge a test, either by arguing that the test is incorrect, or arguing that it does not apply to his implementation. Because of the difficulty and importance of resolving such challenges, a special team of consultants advises the Government on how to resolve the challenges. All challenges are presented to the team without identifying the implementer. Occasionally, a challenge raises a question that has to be resolved by the Language Maintenance Committee.

The possibility of overlooking a test failure was especially likely for the class B tests, which generally contain more than one illegal construct per test.[12] One implementer got in trouble by checking whether the number of error messages issued by the compiler equaled the number of illegal constructs contained in each test; if there was a disagreement, the test output was checked manually. In one test, this implementation issued two error messages for one illegal construct and failed to detect another illegality, but this failure was overlooked since the correct *number* of error messages was produced. The undiagnosed illegal construct was discovered during validation, and even though the compiler error was trivial, no validation certificate was issued.

It can be seen that having more than one illegal construct per test created some difficulties. In particular, the validation process is made more complex because the output of multiple-error tests must be checked manually to see if every error is detected.

Despite these difficulties, I think the decision to allow more than one error per test was correct. In terms of producing tests, it was much easier to produce multiple-error tests than to produce tests containing a single error, since often the same set of declarations could be used to create several error conditions. Since the current set of tests contain 6,540 illegal constructs, if we had allowed only one error per test, instead of having 1,026 class B tests, we would have more than 6,500 such tests (we might have had fewer because of the increased administrative complexity of generating so many individual programs). In addition, having so many more class B tests would certainly lengthen the time required to run all the tests.

Most of the difficulties implementers have in recovering from errors occur for syntax errors. When we noticed this trend, we revised the tests so there would be at most one syntax error per test. With this approach, most implementations were able to diagnose more than one error per test. Since users want good error recovery, having more than one non-syntactic error per test provides some incentive to vendors to improve their error recovery algorithms. On the other hand, since the formal criterion is that an implementation be able to detect any single illegal construct, if a test required more error recovery sophistication than an implementer was willing to provide, the implementer could always request that the test be split. This has seemed to be an acceptable compromise.

8 ACVC Successes and Deficiencies

The ACVC has met DoD's objectives. Although errors exist in validated compilers, and although the test suite is not yet nominally complete, the validation process has

---

[12]After encountering an illegal construct, implementations are not always able to correctly diagnose later illegalities. In such cases, an implementer is allowed to "split" a test into separate subtests containing a single illegal construct per subtest.

undoubtedly improved the conformity of Ada compilers. Particularly notable was the ability to have a substantial number of tests available to check compilers shortly after the official approval of the Standard in February 1983.

There were several unanticipated problems that arose in the ACVC effort:

- inadequate configuration management tools: we did not pay enough attention at the outset to developing or buying tools suitable for managing thousands of individual program files. Since the size of the group writing test programs was small (never more than 5 people), informal methods sufficed to coordinate the writing of tests. But we have increasingly felt the need for automated support tools for:

  - keeping track of change requests and the status of tests

  - prioritizing test development/maintenance work

  - logging and tracking changes made to successive releases

- we have found it impossible to identify all tests that need to be changed when it turns out that an incorrect language interpretation has crept into some tests. Currently, incorrect interpretations are discovered only as implementers report incorrect tests.

- the number of tests needed and the cost to develop each test was significantly underestimated. We initially estimated that 1400 tests would suffice to cover the language with reasonable thoroughness. We currently estimate that about 4200 individual test files are needed to test the language adequately.

- the effort required to write a particular test for a specified test objective was also underestimated. There is effort in the original writing, effort devoted to an independent review, and effort in running the test and making necessary corrections. On the average, 8 person-hours are required for each test. The total cost of developing the test suite is a significant fraction of the cost of developing an Ada compiler.

9 Summary

The major characteristics of the ACVC effort, their impact on Ada's evolution, and their relation to software testing principles and practices are summarized below.

9.1 Test Analysis during Design

The decision to establish an independent test team before Ada's design was even near completion was essential to the success of the ACVC effort and helped considerably in improving the precision of the eventual Standard. The idea of involving quality assurance personnel while software is being designed is not, of course, new, but the benefits of doing so are perhaps not as widely appreciated as they ought to be. The primary benefit stems from having people specifically responsible for ensuring that a specification is free of contradictions and is sufficiently precise to be testable. For a language specification, the testing team checks to be sure a meaning is ascribed to every legal language construct and that there is no doubt about whether certain programs are legal or not. Less attention is paid to whether the ascribed meaning is reasonable, or whether an illegal construct should be made legal (and vice versa). For other kinds of software, similar specification analysis is best performed by people specifically charged with this responsibility.

9.2 Conformity Testing vs. Usability Testing

The ACVC tests focused primarily on evaluating conformity to a specification and did not attempt to evaluate the usability of an Ada implementation. As I have noted, the emphasis on conformity testing was understandable, given DoD's experience with attempts to create standard languages. Nonetheless, it would have been valuable to have had an independent usability test team participate in the evolution of the Ada design, defining potential usability criteria and evaluating the evolving design against potential user needs. A document describing criteria for evaluating Ada compilers for usability in various DoD environments would have provided helpful guidance to compiler developers and could have served a useful role in guiding implementation tradeoffs. Even more valuable would have been tests aimed at evaluating the usability of Ada compilers in a range of typical DoD contexts.

It is certainly difficult to develop sound criteria for evaluating the usability of a software product, but just because it is difficult does not mean the problem should be ignored. A project should identify specific people who will be responsible for developing usability criteria and evaluation tests. These people should participate during the development of a design, but not merely in formal design reviews.

9.3 Test Selection Criteria

The ACVC tests focused on attempting to detect particular kinds of implementation errors. The focus on checking for likely implementation errors encouraged the development of tests that explored the "fringes" of the language, since these were certainly areas where implementers might make a mistake. In terms of creating tests that were easily failed, we probably succeeded (although only the compiler developers can say for sure!).

The ACVC tests are really a kind of acceptance test intended to show that a product meets its specifications sufficiently well that it can be considered ready for use. However, the criteria for deciding what to test did not reflect this viewpoint, in part because it would be hard to decide in advance what combination of language features would be likely to be used, and therefore, should be tested. Nonetheless, we could have asked, for each test, "Would failure of this test be considered sufficiently important to make a compiler unusable?" If the answer was "Definitely not," then perhaps the test should not have been written.

In short, the implications of the test selection criteria on the characteristics of the delivered product should be given extra consideration. It is not enough to focus on correctness of the product as the only criterion guiding test selection. One should consider the relative importance of possible errors as well as the expected usage patterns in devising suitable acceptance tests.

9.4 Cost to Develop Tests

The amount of analytical effort needed to develop good tests is easy to underestimate. It is also easy to assume that inexperienced people can write adequate tests, even given specific test objectives. In our experience, developing good test objectives requires senior analytical skills. If test objectives are being developed while a design evolves, this analytical effort will have the extra benefit of detecting specification errors.

9.5 Impact of the ACVC Effort

The ACVC effort has been successful even though the test suite still has important lapses in coverage. Ada compilers undoubtedly conform more closely to the Standard than would otherwise have been the case. Having the test suite available before compilers were fully developed had its advantages and disadvantages:

- early availability discouraged dialects and encouraged implementers to take the language specification seriously.
- enforcing all aspects of the language before there had been much implementation experience or usage experience meant enforcing some rules that in retrospect required an undue amount of implementation effort given the expected utility to users.
- requiring compilers to pass all the tests delayed the availability of Ada compilers, but had the advantage of preventing the *de facto* use of Ada dialects.

Acknowledgments

Several people provided helpful criticism of an earlier draft of this paper. David Weiss was particularly helpful as the only reviewer who was not already quite familiar with the ACVC. I am also grateful for comments provided by Ron Brender, Ernesto Guerrieri, John Kelly, and Brian Wichmann.

Appendix I

Section 2.4.2 from the *Implementers' Guide*

2.4.2 Based Literals

Semantic Ramifications

S1. The same literal value can be written in different ways in the source text of a program, e.g., 0.01E1 and 0.1; 2#0.01# and 0.25; and 16#0.F# and 4#0.33#. Real literals that do not represent model numbers need not have the same representation in an executed program. For example, 0.1 is not a floating point model number, so different occurrences of 0.1 need not have the same representation, and the representation for 0.1 need not be the same as that for 0.01E1. Of course, any representation of 0.1 must lie within the correct model interval for the floating point type (RM 4.5.7).

S2. The TEXT_IO input routines read real values according to the syntax of a real literal. Since the syntax of a based real literal allows any letter to occur between the sharp signs (#), this means the reading of a real literal cannot stop just because a letter greater than F is found while attempting to read a real literal. See IG 14.3.8/S for further discussion.

S3. RM 2.10/3 allows a based literal to be written with colons replacing each sharp. This affects lexical analysis. Two-character lookahead is needed to process based literals that use colons instead of sharps. Consider:

```
X : INTEGER range 0..2:= 1;
Y : INTEGER range 0..2:10::= 1;
Z : INTEGER range 0..2#10#:= 1;
```

When a digit is followed by a colon, it is not possible to tell if the sequence is the start of a based number or if the colon is the beginning of the := compound delimiter. This ambiguity can only be resolved by looking at the character that follows the colon.

Legality Rules

L1. The base value must be greater than or equal to 2 and less than or equal to 16 (RM 2.4.2/1).

L2. The letters and digits occurring between sharps must have a value less than the value of the base (RM 2.4.2/4).

L3. The exponent for an integer based literal must not contain a minus sign (RM 2.4.1/4).

L4. An integer based literal must not contain a point (RM 2.4/1).

L5. If one sharp is replaced with a colon, the other must also be replaced (RM 2.10/3).

L6. No based literal can contain more characters than the maximum input line length permitted by an implementation (RM 2.2/3).

Test Objectives

T1. Check that a based literal always yields a nonnegative value.

T2. Check that nonconsecutive embedded underscores are permitted in every part of a based literal, and do not affect the value.

T3. Check that based literals with bases 2 through 16 all yield correct values.

T4. Check that the digits and extended digits of a based literal are within the correct range for the number's base.

T5. Check that negative exponents are forbidden in integer based literals. *Implementation Guideline:* Use an exponent of -0 in one case.

Check that negative exponents are allowed in real based literals (implicitly checked by IG 2.4.1/T13).

T6. Check that the base must not be less than 2 or greater than 16.

T7. Check that letters in a based literal may appear in upper or lower case.

T8. Check that underscores may not be adjacent to the '#' (see IG 2.4/T1).

T9. Check that 'E' and 'e' may appear in based literals (see IG 2.4/T2).

T10. Check that leading zeros in based literals are ignored (see IG 2.4/T3).
Check that based literals can be as long as the maximum input line length (see IG 2.4/T3).

T11. Check that two-character lookahead is used for based literals having colons in place of sharps.

Appendix II

Example of a Class A Test

The REPORT package contains routines that are used by all executable tests. The call to the TEST procedure prints out information about the test when it is executed. The call to the RESULT procedure prints out a passed/failed message. In the case of class A tests, if the test compiles successfully, it should also execute successfully. The result of executing this test is given in Appendix IX.

```
-- CHECK THAT KEYWORDS OF OTHER LANGUAGES WHICH ARE NOT KEYWORDS
--     IN ADA CAN BE USED AS VARIABLE NAMES.

-- CHECKS KEYWORDS BEGINNING WITH A - B.

-- DCB 6/6/80
-- JRK 7/3/80
-- SPS 10/21/82
-- SPS 2/11/83

WITH REPORT;
PROCEDURE A29002A IS

     USE REPORT;

  -- ABORT           : INTEGER;   -- J73
  -- ABS             : INTEGER;   -- ALGOL68, J3B, J73, J73/I, PL/I,
                                  -- TACPOL
  -- ACCEPT          : INTEGER;   -- COBOL
  -- ACCESS          : INTEGER;   -- COBOL
     ACOS            : INTEGER;   -- TACPOL
     ACTIVATE        : INTEGER;   -- SPL/I
     ADD             : INTEGER;   -- COBOL, PL/I
     ADDR            : INTEGER;   -- PL/I
     ADDREL          : INTEGER;   -- PL/I
     ADDRESSSIZE     : INTEGER;   -- J73/I
     ADVANCING       : INTEGER;   -- COBOL
     AFTER           : INTEGER;   -- COBOL, PL/I
     ALIGNED         : INTEGER;   -- TACPOL, PL/I
  -- ALL             : INTEGER;   -- COBOL, J73/I
     ALLOC           : INTEGER;   -- SPL/I
     ALLOCATE        : INTEGER;   -- PLC, PL/I
     ALLOCATION      : INTEGER;   -- PL/I
     ALPHABETIC      : INTEGER;   -- COBOL
     ALSO            : INTEGER;   -- COBOL
     ALTER           : INTEGER;   -- COBOL
     ALTERNATE       : INTEGER;   -- COBOL
  -- AND             : INTEGER;   -- ALGOL68, CMS-2, COBOL, J3B, J73,
                                  -- J73/I, PASCAL, SPL/I, TACPOL
-- other declarations omitted here
BEGIN
     TEST ("A29002A", "NO ADDITIONAL RESERVED WORDS");
     RESULT;
END A29002A;
```

Appendix III

Example of a Class B Test

Class B tests contain illegal constructs. The line on which the illegal construct occurs is marked with a comment that starts **ERROR:**. This test is one of the tests written for test objective T4 for the section of the *Implementers' Guide* given in Appendix I.

```
-- B24204A-AB.ADA

-- CHECK THAT THE DIGITS AND EXTENDED_DIGITS OF AN INTEGER LITERAL
-- ARE WITHIN THE CORRECT RANGE FOR THE NUMBER'S BASE.

-- JRK 12/12/79
-- JRK 10/27/80
-- JWC 6/28/85    RENAMED FROM B24104A.ADA

PROCEDURE B24204A IS

        I  : INTEGER;

BEGIN

        I := 2#2#;              -- ERROR: 2#2#
        I := 3#3#;              -- ERROR: 3#3#
        I := 4#4#;              -- ERROR: 4#4#
        I := 5#5#;              -- ERROR: 5#5#
        I := 6#6#;              -- ERROR: 6#6#
        I := 7#7#;              -- ERROR: 7#7#
        I := 8#8#;              -- ERROR: 8#8#
        I := 9#9#;              -- ERROR: 9#9#
        I := 10#A#;             -- ERROR: 10#A#
        I := 11#B#;             -- ERROR: 11#B#
        I := 12#C#;             -- ERROR: 12#C#
        I := 13#D#;             -- ERROR: 13#D#
        I := 14#E#;             -- ERROR: 14#E#
        I := 15#F#;             -- ERROR: 15#F#
        I := 16#G#;             -- ERROR: 16#G#

        I := 3A;                -- ERROR: 3A
        I := 0A#2#;             -- ERROR: 0A#2#
        I := 2E1A;              -- ERROR: 2E1A

END B24204A;
```

Appendix IV

Example of a Class C Test

Class C tests are executable and self-checking. See test objective T2 in the example from the *Implementers' Guide*, Appendix I. The call to the TEST procedure prints out information about the test when it is executed. The call to the RESULT procedure prints out a passed/failed message (see Appendix IX).

```
-- C24202A-AB.ADA

-- CHECK THAT NON-CONSECUTIVE UNDERSCORES ARE PERMITTED
-- IN EVERY PART OF A BASED INTEGER LITERAL.

-- DCB 1/24/80
-- JRK 10/27/80
-- TBN 10/17/85     RENAMED FROM C24102A.ADA AND FIXED LINE LENGTH.

WITH REPORT;
PROCEDURE C24202A IS

     USE REPORT;

     I1, I2, I3, I4 : INTEGER;

BEGIN
     TEST("C24202A", "UNDERSCORES ALLOWED IN INTEGER LITERALS");

     I1 := 12_3;
     I2 := 1_6#D#;
     I3 := 2#1011_0101#;
     I4 := 16#D#E0_1;

     IF I1 = 123 AND I2 = 16#D# AND I3 = 2#10110101# AND
        I4 = 16#D#E01 THEN
          NULL;
     ELSE
          FAILED("UNDERSCORES IN INTEGER LITERALS " &
                 "NOT HANDLED CORRECTLY");
     END IF;

     RESULT;
END C24202A;
```

Appendix V

Example of a Class D Test

Class D tests check capacity limits of implementations. This particular test checks whether 64-bit integer literals are supported correctly in computations performed at compile-time.

```
-- D4A002B.ADA

-- LARGER LITERALS IN NUMBER DECLARATIONS, BUT WITH RESULTING
-- SMALLER VALUE OBTAINED BY SUBTRACTION. THIS TEST LIMITS VALUES
-- TO 64 BINARY PLACES.

-- BAW 29 SEPT 80

WITH REPORT;
PROCEDURE D4A002B IS

    USE REPORT;

    X : CONSTANT := 4123456789012345678 - 4123456789012345679;
    Y : CONSTANT := 4 * (10 ** 18) - 3999999999999999999;
    Z : CONSTANT := (1024 ** 6) - (2 ** 60);
    D : CONSTANT := 9_223_372_036_854_775_807/20_303_320_287_433;
    E : CONSTANT := 36_028_790_976_242_271 REM 17_600_175_361;
    F : CONSTANT := ( - 2 ** 51 ) MOD ( - 131_071 );

BEGIN TEST("D4A002B","LARGE INTEGER RANGE (WITH CANCELLATION) IN " &
          "NUMBER DECLARATIONS; LONGEST INTEGER IS 64 BITS ");

    IF X /= -1 OR Y /= 1 OR Z /= 0
       OR D /= 454_279 OR E /= 1 OR F /= -1
    THEN FAILED("EXPRESSIONS WITH A LARGE INTEGER RANGE (WITH " &
                "CANCELLATION) ARE NOT EXACT ");
    END IF;

    RESULT;

END D4A002B;
```

Appendix VI

Example of a Class E Test

The pass/fail criterion for a class E test is unique to each test and is given in the test header.

```
-- EA1003B-B.ADA

-- CHECK WHETHER MORE THAN ONE COMPLETELY INDEPENDENT COMPILATION
-- UNIT CAN BE SUBMITTED IN A SINGLE FILE AND WHETHER ERRORS IN
-- ONE INDEPENDENT UNIT AFFECT LATER INDEPENDENT UNITS IN THE FILE.

-- AN IMPLEMENTATION IS ALLOWED BY AI-00255 TO REJECT LEGAL
-- UNITS IN THIS COMPILATION FILE IN ADDITION TO THE ILLEGAL ONE,
-- EA1003B_PKG, IN WHICH CASE THE TEST SHOULD NOT BE LINKABLE.

-- JRK 5/12/81
-- JRK 11/21/85  RENAMED FROM CA1003B-AB.ADA; REVISED ACCORDING TO
--               AI-00255.


PROCEDURE EA1003B_P (I : IN OUT INTEGER) IS
BEGIN
     I := I + 1;
END EA1003B_P;


PACKAGE EA1003B_PKG IS
     I : T;       -- ERROR: UNDEFINED TYPE NAME.
END EA1003B_PKG;


FUNCTION EA1003B_F (I : INTEGER) RETURN INTEGER IS
BEGIN
     RETURN -I;
END EA1003B_F;

WITH REPORT, EA1003B_P, EA1003B_F;
USE REPORT;
PROCEDURE EA1003B IS

     I : INTEGER := IDENT_INT (0);

BEGIN

     TEST ("EA1003B", "INDEPENDENT UNITS IN A SINGLE FILE " &
          "ARE NOT AFFECTED BY ERRORS IN OTHER UNITS");

     EA1003B_P (I);
     IF I /= 1 THEN
          FAILED ("INDEPENDENT PROCEDURE NOT INVOKED");
     END IF;
```

```
    IF EA1003B_F (IDENT_INT(5)) /= -5 THEN
          FAILED ("INDEPENDENT FUNCTION NOT INVOKED");
    END IF;

    RESULT;
END EA1003B;
```

Appendix VII

Result of Compiling the Class B Test

The class B test given in Appendix III was compiled with the NYU Ada/ED compiler, with the result indicated below. This test was passed, since each line that contains an error was diagnosed by the implementation. The adequacy of the error diagnostic is not evaluated by the ACVC. It is sufficient if the implementation correctly rejects all and only those lines that contain illegal constructs.

```
NYU Ada/ED-C 1.7  VAX BSD 4.2      Mon  30 Jun 1986  16:49:23  PAGE     1


   1:    -- B24204A-AB.ADA
   2:
   3:    -- CHECK THAT THE DIGITS AND EXTENDED_DIGITS OF AN INTEGER LITERAL
   4:    -- ARE WITHIN THE CORRECT RANGE FOR THE NUMBER'S BASE.
   5:
   6:    -- JRK 12/12/79
   7:    -- JRK 10/27/80
   8:    -- JWC 6/28/85   RENAMED FROM B24104A.ADA
   9:
  10:    PROCEDURE B24204A IS
  11:
  12:          I  : INTEGER;
  13:
  14:    BEGIN
  15:
  16:          I := 2#2#;              -- ERROR: 2#2#
                      ^
*** ERROR: Invalid based-number digit

  17:          I := 3#3#;              -- ERROR: 3#3#
                      ^
*** ERROR: Invalid based-number digit

  18:          I := 4#4#;              -- ERROR: 4#4#
                      ^
*** ERROR: Invalid based-number digit

  19:          I := 5#5#;              -- ERROR: 5#5#
                      ^
*** ERROR: Invalid based-number digit

  20:          I := 6#6#;              -- ERROR: 6#6#
                      ^
*** ERROR: Invalid based-number digit

  21:          I := 7#7#;              -- ERROR: 7#7#
                      ^
*** ERROR: Invalid based-number digit
```

```
  22:            I := 8#8#;                  -- ERROR: 8#8#
                    ^
*** ERROR: Invalid based-number digit
  23:            I := 9#9#;                  -- ERROR: 9#9#
                    ^
*** ERROR: Invalid based-number digit

  24:            I := 10#A#;                 -- ERROR: 10#A#
                     ^
*** ERROR: Invalid based-number digit

  25:            I := 11#B#;                 -- ERROR: 11#B#
                     ^
*** ERROR: Invalid based-number digit

  26:            I := 12#C#;                 -- ERROR: 12#C#
                     ^
*** ERROR: Invalid based-number digit

  27:            I := 13#D#;                 -- ERROR: 13#D#
                     ^
*** ERROR: Invalid based-number digit

  28:            I := 14#E#;                 -- ERROR: 14#E#
                     ^
*** ERROR: Invalid based-number digit

  29:            I := 15#F#;                 -- ERROR: 15#F#
                     ^
*** ERROR: Invalid based-number digit

  30:            I := 16#G#;                 -- ERROR: 16#G#
                     ^
*** ERROR: Incomplete based number
                   ^
*** ERROR: Expect '#' after last digit
                      <->
*** ERROR: Number should be separated from adjacent identifier
                       ^
*** ERROR: Bad character in file ignored: #
                       <->
*** ERROR: ";" expected after this token
                     ^
*** ERROR: identifier undeclared or not visible G (RM 3.1)

  31:
  32:            I := 3A;                    -- ERROR: 3A
                     ^
*** ERROR: Number should be separated from adjacent identifier
                    ^
*** ERROR: ";" expected after this token
```

NYU Ada/ED-C 1.7   VAX BSD 4.2   Mon 30 Jun 1986 16:49:23 PAGE   3
                              ^
*** ERROR: identifier undeclared or not visible A (RM 3.1)
   33:            I := 0A#2#;   -- ERROR: 0A#2#
                              ^
*** ERROR: Number should be separated from adjacent identifier
                         ^
*** ERROR: Bad character in file ignored: #
                           ^
*** ERROR: Incomplete based number
                            ^
*** ERROR: Expect '#' after last digit
                          ^
*** ERROR: One of { ** REM MOD * / - + & IN <= >= /= > = < OR AND XOR }
   expected instead of "A"
   34:            I := 2E1A;   -- ERROR: 2E1A
                       <->
*** ERROR: Number should be separated from adjacent identifier
                       <->
*** ERROR: ";" expected after this token

   35:
   36:    END B24204A;
   37:

          30 errors detected

225

Appendix VIII

Result of Compiling the Class E Test

NYU Ada/ED-C 1.7 VAX BSD 4.2   Mon 30 Jun 1986 16:51:28 PAGE   1

```
 1:    -- EA1003B-B.ADA
 2:
 3:    -- CHECK WHETHER MORE THAN ONE COMPLETELY INDEPENDENT COMPILATION
 4:    -- UNIT CAN BE SUBMITTED IN A SINGLE FILE AND WHETHER ERRORS IN
 5:    -- ONE INDEPENDENT UNIT AFFECT LATER INDEPENDENT UNITS IN THE FILE.
 6:
 7:    -- AN IMPLEMENTATION IS ALLOWED BY AI-00255 TO REJECT LEGAL
 8:    -- UNITS IN THIS COMPILATION FILE IN ADDITION TO THE ILLEGAL ONE,
 9:    -- EA1003B_PKG, IN WHICH CASE THE TEST SHOULD NOT BE LINKABLE.
10:
11:    -- JRK 5/12/81
12:    -- JRK 11/21/85  RENAMED FROM CA1003B-AB.ADA; REVISED ACCORDING TO
13:    --              AI-00255.
14:
15:
16:    PROCEDURE EA1003B_P (I : IN OUT INTEGER) IS
17:    BEGIN
18:        I := I + 1;
19:    END EA1003B_P;
20:
21:
22:    PACKAGE EA1003B_PKG IS
23:        I : T;      -- ERROR: UNDEFINED TYPE NAME.
                       ^
*** ERROR: identifier undeclared or not visible T (RM 3.1)

24:    END EA1003B_PKG;
25:
26:
27:    FUNCTION EA1003B_F (I : INTEGER) RETURN INTEGER IS
28:    BEGIN
29:        RETURN -I;
30:    END EA1003B_F;
31:
32:
33:    WITH REPORT, EA1003B_P, EA1003B_F;
34:    USE REPORT;
35:
36:    PROCEDURE EA1003B IS

         Between line 33 column 1 and line 36 column 9
*** ERROR: Unknown unit in with clause: EA1003B_F (RM 10.1.1)
```

```
37:
38:          I : INTEGER := IDENT_INT (0);
39:
40:    BEGIN
41:          TEST ("EA1003B", "INDEPENDENT UNITS IN A SINGLE FILE " &
42:                 "ARE NOT AFFECTED BY ERRORS IN OTHER UNITS");
43:
44:          EA1003B_P (I);
45:          IF I /= 1 THEN
46:                FAILED ("INDEPENDENT PROCEDURE NOT INVOKED");
47:          END IF;
48:
49:          IF EA1003B_F (IDENT_INT(5)) /= -5 THEN
                 <------->
*** ERROR: premature use of EA1003B_F (RM 8.3)

50:                FAILED ("INDEPENDENT FUNCTION NOT INVOKED");
51:          END IF;
52:
53:          RESULT;
54:    END EA1003B;
55:
```

        3 errors detected

This test is passed. The illegality on line 23 is detected, as required. In addition, as explained in lines 7-9, an implementation need not add any of the compilation units to the library. In particular, this implementation does not add EA1003B_F to the library. Because this unit is not added to the library, the WITH clause in line 33 is rejected; unit EA1003B_F cannot be found. This leads to a later error in line 49.

Appendix IX

Results of Executing Tests

This Appendix contains the results of executing the class A, C, and D tests contained in preceding Appendices. In addition, to show how failed tests are indicated, the result of executing test CZ1103A-B.ADA is included: informational comments are preceded by a single dash, and messages indicating a failure are preceded by a single asterisk. Since at least one failure message was printed, the final result is highlighted with asterisks to show that failure occurred. (In this test, the failure message is expected, since the test is checking that failure will indeed be reported when it has occurred.) The full text of test CZ1103A is in the next Appendix.

```
---- A29002A NO ADDITIONAL RESERVED WORDS.
==== A29002A PASSED ====================.

---- C24202A UNDERSCORES ALLOWED IN INTEGER LITERALS.
==== C24202A PASSED ====================.

---- D4A002B LARGE INTEGER RANGE (WITH CANCELLATION) IN NUMBER
              DECLARATIONS; LONGEST INTEGER IS 64 BITS .
==== D4A002B PASSED ====================.

---- CZ1103A CHECK THAT PROCEDURE CHECK_FILE WORKS.
   - CZ1103A BEGIN TEST WITH AN EMPTY FILE.
   - CZ1103A BEGIN TEST WITH A FILE WITH BLANK LINES.
   - CZ1103A BEGIN TEST WITH A FILE WITH BLANK LINES AND PAGES.
   - CZ1103A BEGIN TEST WITH A FILE WITH TRAILING BLANKS.
   - CZ1103A FROM CHECK_FILE: THIS IMPLEMENTATION PADS LINES WITH
              BLANKS.
   - CZ1103A BEGIN TEST WITH A FILE WITHOUT TRAILING BLANKS.
   - CZ1103A BEGIN TEST WITH A FILE WITH AN END OF LINE ERROR.
   * CZ1103A FROM CHECK_FILE: END OF LINE EXPECTED - E ENCOUNTERED.
   - CZ1103A FROM CHECK_FILE: LAST CHARACTER IN FOLLOWING STRING
              REVEALED ERROR: THIS LINE WILL CONTAIN AN #.
   - CZ1103A BEGIN TEST WITH FILE WITH END OF PAGE ERROR.
   * CZ1103A FROM CHECK_FILE: END_OF_PAGE NOT WHERE EXPECTED.
   - CZ1103A FROM CHECK_FILE: LAST CHARACTER IN FOLLOWING STRING
              REVEALED ERROR: THIS LINE WILL CONTAIN AN @.
   - CZ1103A BEGIN TEST WITH FILE WITH END OF FILE ERROR.
   * CZ1103A FROM CHECK_FILE: END_OF_FILE NOT WHERE EXPECTED.
   - CZ1103A FROM CHECK_FILE: LAST CHARACTER IN FOLLOWING STRING
              REVEALED ERROR: THIS LINE WILL CONTAIN AN %.
   - CZ1103A BEGIN TEST WITH FILE WITH INCORRECT DATA.
   * CZ1103A FROM CHECK_FILE: FILE DOES NOT CONTAIN CORRECT OUTPUT -
              EXPECTED C - GOT I.
   - CZ1103A FROM CHECK_FILE: LAST CHARACTER IN FOLLOWING STRING
              REVEALED ERROR: LINE WITH C.
**** CZ1103A FAILED ********************.
```

Appendix X
Test CZ1103A

```
-- CZ1103A-B.ADA

-- CHECK THAT THE PROCEDURE CHECK_FILE WORKS CORRECTLY, IN PARTICULAR,
-- THAT IT WILL REPORT INCORRECT FILE CONTENTS AS TEST FAILURES.

-- THIS TEST INTENTIONALLY CONTAINS MISMATCHES BETWEEN FILE CONTENTS AND
-- THE 'CONTENTS' STRING PARAMETER OF PROCEDURE CHECK_FILE.

-- AN IMPLEMENTATION PASSES THIS TEST IF TEST EXECUTION REPORTS THE
-- FOLLOWING FOUR FAILURES (AND A FINAL 'FAILED' RESULT) AND REPORTS NO
-- OTHER FAILURES:
--       * CZ1103A FROM CHECK_FILE: END OF LINE EXPECTED - E ENCOUNTERED.
--       * CZ1103A FROM CHECK_FILE: END_OF_PAGE NOT WHERE EXPECTED.
--       * CZ1103A FROM CHECK_FILE: END_OF_FILE NOT WHERE EXPECTED.
--       * CZ1103A FROM CHECK_FILE: FILE DOES NOT CONTAIN CORRECT OUTPUT -
--              EXPECTED C - GOT I.

-- SPS 12/9/82
-- JRK 11/18/85  ADDED COMMENTS ABOUT PASS/FAIL CRITERIA.

WITH REPORT; USE REPORT;
WITH TEXT_IO; USE TEXT_IO;
WITH CHECK_FILE;

PROCEDURE CZ1103A IS

     NULL_FILE : FILE_TYPE;
     FILE_WITH_BLANK_LINES : FILE_TYPE;
     FILE_WITH_BLANK_PAGES : FILE_TYPE;
     FILE_WITH_TRAILING_BLANKS : FILE_TYPE;
     FILE_WITHOUT_TRAILING_BLANKS : FILE_TYPE;
     FILE_WITH_END_OF_LINE_ERROR : FILE_TYPE;
     FILE_WITH_END_OF_PAGE_ERROR : FILE_TYPE;
     FILE_WITH_END_OF_FILE_ERROR : FILE_TYPE;
     FILE_WITH_DATA_ERROR : FILE_TYPE;

BEGIN

     TEST ("CZ1103A", "CHECK THAT PROCEDURE CHECK_FILE WORKS");

     BEGIN
          COMMENT ("BEGIN TEST WITH AN EMPTY FILE");
          CREATE (NULL_FILE, OUT_FILE);
          CHECK_FILE (NULL_FILE, "#@%");
          CLOSE (NULL_FILE);
     EXCEPTION
          WHEN OTHERS =>
               FAILED ("TEST WITH EMPTY FILE INCOMPLETE");
     END;
```

```
-- THIS SECTION TESTS CHECK_FILE WITH A FILE WITH BLANK LINES.

    BEGIN
        COMMENT ("BEGIN TEST WITH A FILE WITH BLANK LINES");
        CREATE (FILE_WITH_BLANK_LINES, OUT_FILE);
        NEW_LINE (FILE_WITH_BLANK_LINES, 20);
        CHECK_FILE (FILE_WITH_BLANK_LINES, "####################@%");
        CLOSE (FILE_WITH_BLANK_LINES);
    EXCEPTION
        WHEN OTHERS =>
            FAILED ("TEST WITH FILE WITH BLANK LINES INCOMPLETE");
    END;

-- THIS SECTION TESTS CHECK_FILE WITH A FILE WITH BLANK LINES AND PAGES.

    BEGIN
        COMMENT ("BEGIN TEST WITH A FILE WITH BLANK LINES " &
                 "AND PAGES");
        CREATE (FILE_WITH_BLANK_PAGES, OUT_FILE);
        NEW_LINE (FILE_WITH_BLANK_PAGES, 3);
        NEW_PAGE (FILE_WITH_BLANK_PAGES);
        NEW_LINE (FILE_WITH_BLANK_PAGES, 2);
        NEW_PAGE (FILE_WITH_BLANK_PAGES);
        NEW_PAGE (FILE_WITH_BLANK_PAGES);
        CHECK_FILE (FILE_WITH_BLANK_PAGES, "###@##@#@%");
        CLOSE (FILE_WITH_BLANK_PAGES);
    EXCEPTION
        WHEN OTHERS =>
            FAILED ("TEST WITH FILE WITH BLANK PAGES INCOMPLETE");
    END;

-- THIS SECTION TESTS CHECK_FILE WITH A FILE WITH TRAILING BLANKS.

    BEGIN
        COMMENT ("BEGIN TEST WITH A FILE WITH TRAILING BLANKS");
        CREATE (FILE_WITH_TRAILING_BLANKS, OUT_FILE);
        FOR I IN 1 .. 3 LOOP
            PUT_LINE (FILE_WITH_TRAILING_BLANKS,
                  "LINE WITH TRAILING BLANKS     ");
        END LOOP;
        CHECK_FILE(FILE_WITH_TRAILING_BLANKS, "LINE WITH TRAILING" &
                   " BLANKS#LINE WITH TRAILING BLANKS#LINE" &
                   " WITH TRAILING BLANKS#@@%");
        CLOSE (FILE_WITH_TRAILING_BLANKS);
    EXCEPTION
        WHEN OTHERS =>
            FAILED ("TEST WITH FILE WITH TRAILING BLANKS " &
                    "INCOMPLETE");
    END;
```

```
-- THIS SECTION TESTS CHECK_FILE WITH A FILE WITHOUT TRAILING BLANKS.

     BEGIN
          COMMENT ("BEGIN TEST WITH A FILE WITHOUT TRAILING BLANKS");
          CREATE (FILE_WITHOUT_TRAILING_BLANKS, OUT_FILE);
          FOR I IN 1 .. 3 LOOP
               PUT_LINE (FILE_WITHOUT_TRAILING_BLANKS,
                         "LINE WITHOUT TRAILING BLANKS");
          END LOOP;
          CHECK_FILE(FILE_WITHOUT_TRAILING_BLANKS, "LINE WITHOUT " &
                    "TRAILING BLANKS#LINE WITHOUT TRAILING BLANKS#" &
                    "LINE WITHOUT TRAILING BLANKS#@%");
          CLOSE (FILE_WITHOUT_TRAILING_BLANKS);
     EXCEPTION
          WHEN OTHERS =>
               FAILED ("TEST WITH FILE WITHOUT TRAILING BLANKS " &
                       "INCOMPLETE");
     END;

-- THIS SECTION TESTS CHECK_FILE WITH A FILE WITH AN END OF LINE ERROR.

     BEGIN
          COMMENT ("BEGIN TEST WITH A FILE WITH AN END OF LINE ERROR");
          CREATE (FILE_WITH_END_OF_LINE_ERROR, OUT_FILE);
          PUT_LINE (FILE_WITH_END_OF_LINE_ERROR, "THIS LINE WILL " &
                    "CONTAIN AN END OF LINE IN THE WRONG PLACE");
          CHECK_FILE (FILE_WITH_END_OF_LINE_ERROR, "THIS LINE WILL " &
               "CONTAIN AN # IN THE WRONG PLACE#@%");
          CLOSE (FILE_WITH_END_OF_LINE_ERROR);
     EXCEPTION
          WHEN OTHERS =>
               FAILED ("TEST WITH END_OF_LINE ERROR INCOMPLETE");
     END;

-- THIS SECTION TESTS CHECK_FILE WITH A FILE WITH AN END OF PAGE ERROR.

     BEGIN
          COMMENT ("BEGIN TEST WITH FILE WITH END OF PAGE ERROR");
          CREATE (FILE_WITH_END_OF_PAGE_ERROR, OUT_FILE);
          PUT_LINE (FILE_WITH_END_OF_PAGE_ERROR, "THIS LINE WILL " &
                    "CONTAIN AN END OF PAGE IN THE WRONG PLACE");
          CHECK_FILE (FILE_WITH_END_OF_PAGE_ERROR, "THIS LINE WILL " &
               "CONTAIN AN @ IN THE WRONG PLACE#@%");
          CLOSE (FILE_WITH_END_OF_PAGE_ERROR);
     EXCEPTION
          WHEN OTHERS =>
               FAILED ("TEST WITH END_OF_PAGE ERROR INCOMPLETE");
     END;
```

```
-- THIS SECTION TESTS CHECK_FILE WITH A FILE WITH AN END OF FILE ERROR.

      BEGIN
            COMMENT ("BEGIN TEST WITH FILE WITH END OF FILE ERROR");
            CREATE (FILE_WITH_END_OF_FILE_ERROR, OUT_FILE);
            PUT_LINE (FILE_WITH_END_OF_FILE_ERROR, "THIS LINE WILL " &
                      "CONTAIN AN END OF FILE IN THE WRONG PLACE");
            CHECK_FILE (FILE_WITH_END_OF_FILE_ERROR, "THIS LINE WILL " &
                      "CONTAIN AN % IN THE WRONG PLACE#@%");
            CLOSE (FILE_WITH_END_OF_FILE_ERROR);
      EXCEPTION
            WHEN OTHERS =>
                  FAILED ("TEST WITH END_OF_FILE ERROR INCOMPLETE");
      END;

-- THIS SECTION TESTS CHECK_FILE WITH A FILE WITH INCORRECT DATA.

      BEGIN
            COMMENT ("BEGIN TEST WITH FILE WITH INCORRECT DATA");
            CREATE (FILE_WITH_DATA_ERROR, OUT_FILE);
            PUT_LINE (FILE_WITH_DATA_ERROR, "LINE WITH INCORRECT " &
                      "DATA");
            CHECK_FILE (FILE_WITH_DATA_ERROR, "LINE WITH CORRECT " &
                      "DATA#@%");
            CLOSE (FILE_WITH_DATA_ERROR);
      EXCEPTION
            WHEN OTHERS =>
                  FAILED ("TEST WITH INCORRECT DATA INCOMPLETE");
      END;

      RESULT;

END CZ1103A;
```

References

[Ada 80] *Reference Manual for the Ada Programming Language*
MIL-STD-1815 edition, 1980.

[Ada 82] *Reference Manual for the Ada Programming Language*
Draft, July 1982 edition, 1982.

[Ada 83] *Reference Manual for the Ada Programming Language*
ANSI/MIL-STD-1815A-1983 edition, 1983.

[AdaIG 80] Goodenough, J. B.
Ada Compiler Validation Implementers' Guide.
Technical Report, SofTech, Inc., October, 1980.

[AdaIG 86] Goodenough, J. B.
Ada Compiler Validation Implementers' Guide.
Technical Report, SofTech, Inc., December, 1986.

[Carlson 81] Carlson, W. E.
Ada: A promising beginning.
COMPUTER 14(6):13-15, June, 1981.

[Currit 86] Currit, P. A., Dyer, M., and Mills, H. D.
Certifying the reliability of software.
IEEE Transactions on Software Engineering SE-12(1):3-11, January,
1986.

[Goodenough 81] Goodenough, J. B.
The Ada compiler validation capability.
COMPUTER 14(6):57-64, June, 1981.

[Hoyt 77] Hoyt, P. M.
The Navy FORTRAN validation system.
In *AFIPS Conference Proceedings 1977 National Computer
Conference*, pages 529-537. 1977.

[Oliver 79] Oliver, P.
Experiences in building and using compiler validation systems.
In *AFIPS Conference Proceedings 1979 National Computer
Conference*, pages 1051-1057. 1979.

[Wichmann 83] Wichmann, B. A. and Ciechanowicz, Z. J. (editors).
Pascal Compiler Validation.
John Wiley & Sons, 1983.

The Software Engineering Institute
at Carnegie Mellon University

A. N. Habermann
Carnegie Mellon University

Abstract

More and more organizations depend on software for part of their activities. Software suppliers have great difficulty satisfying the increasing demand. The US Department of Defense, being a large user of software, has taken the initiative to push for increased programmers' productivity by improving the software production process. The first phase of the DoD's Software Initiative resulted in the design of the Ada language. The second phase, in which the emphasis has shifted from programs to systems and management, resulted in a variety of activities of which the Software Engineering Institute is one. The objective of the SEI is technology transition in the area of software development support. The SEI has selected a number of areas and topics of interest and has planned a sequence of phases in which more advanced technology is introduced. The SEI has started a number of projects to explore technology that is ready for transition. Several of these projects are related to the Ada Language.

1. Introduction

Our society is becoming more and more dependent on computer software. The production of software systems has increased by an order of magnitude but the demand has increased even more and keeps rising [7]. This situation puts severe pressure on the software manufacturing community which is plagued by a severe shortage of well-educated and well-trained software engineers. An additional aggravating factor is the fact that customers demand much higher standards of quality and reliability than ten years ago.

It is generally recognized that the software production practices have not kept up with the rising demands for quantity, size and quality. Many software systems are produced with technology that dates back to the sixties and early seventies. Systems are often written in special purpose programming languages, depend on peculiar hardware features and make little use of software engineering principles such as information hiding and data abstraction [1]. Such systems are not portable, are error-prone when modified and are critically dependent on what their implementors can remember.

The US Department of Defense, a major consumer of software systems, has taken the initiative to improve the software production process [2]. The DoD Software Initiative started in the mid-seventies with the design of the Ada programming language [3], which largely solves the portability problem. The language also provides direct support for software engineering principles through its package mechanism.

The Ada language was primarily designed for supporting the individual programmer in writing programs that depend on program modules written by others [4]. In contrast to Ada's elaborate support for what is generally known as "programming-in-the-small", the language provides only limited support for "programming-in-the-large" and practically no support for "programming-in-the-many" [5]. Programming-in-the-large is the term for the integration of program modules into systems and for matters such as version control and configuration management. Programming-in-the-many refers to the fact that software systems are typically generated by teams of people, which necessitates task coordination and project management.

It has frequently been stated that improving the software production process is largely a matter of developing good methods for programming-in-the-large and for programming-in-the-many. This insight made the DoD follow up on its Ada initiative with various efforts to stimulate the use of good software engineering principles, methods and tools in practice. One of these efforts gave rise to the Software Engineering Institute at Carnegie Mellon University.

The main objective of the Software Engineering Institute is technology transition. Its main occupation is not basic research, but the transition of existing technology into routine practice. The SEI will have achieved its goal if it succeeds in reducing the time lag between the development of software production methods and tools in the research environment and their application in practice. Thus, the main task of the Software Engineering Institute is to find out what kind of promising methods and tools exist in the research environment, to understand the production environment and explore which of the methods and tools are applicable and, last but not least, to demonstrate their feasibility through the construction of prototype programming environments and through the training of personnel.

The Software Engineering Institute is to a limited extent involved in basic research and education. Its role in these areas is primarily one of stimulating research and education in software engineering by inviting visitors for extended periods of time and by coordinating projects undertaken with affiliates from government, industry and academia. Although these activities do not constitute the major occupation of the SEI, they do amount to an effort that is comparable in size to that of an average computer science department in the US.

The SEI has established a five year plan and has started a number of projects that address the issues described in that plan. The following sections present the essence of that plan and describe in a little more detail the state of Ada-related projects of the SEI. The basic idea behind the five year plan are described in an earlier paper written when the projects were just started [6]. That paper also describes the SEI's organization and its project planning procedures.

2. The SEI's Basic Themes

The Software Engineering Institute will focus on issues involving the production and maintenance of large software systems. The SEI has decided not to promote a particular software development methodology or philosophy, but to review the various issues involving the production process and product quality. The SEI's major task is to demonstrate the available software development support technology and show how it can be applied. A major decision has been to achieve improvement of software production and maintenance by becoming more technology-intensive instead of labor-intensive. We at the SEI believe that production and maintenance can be improved considerably if they are directly supported by the programming environment that provides the software development tools instead of having to rely on labor-intensive management procedures and tenuous conventions applied by humans. One of the major benefits of the technology-intensive approach is that the programming environment can apply consistency checks and can enforce management rules which is a time-consuming task for humans and hard to do consistently, instantly and accurately all the time.

For the initial five year period of its existence, the SEI has chosen three areas and three topics which together constitute the six themes that serve as guidelines for selecting projects. Subdivided into areas and topics, the six themes are:

Areas of Interest	Topics of Interest
1. Technology Identification and Assessment	1. Reasoning about the Software Production Process
2. Nature of the Transition Process	2. Tools and Environments for Software Development
3. Education and Training	3. Reusability and Automation

2.1. Areas of Interest

All three areas concern the essence of the SEI: technology transition. The first area includes the task of finding promising software development technology, of exploring who may benefit from it and of showing how it can be applied. We emphasize that the SEI's task is *assessment* and not *evaluation*. The SEI has the task of showing which existing technology can be applied when and where, but not of publishing consumer reports of commercial products. The motivation for the second area of interest, the Nature of the Transition Process, is that there is ample evidence that new technology is often not applied because of reasons that have nothing to do with the merits of the technology itself, but with the organization of the site of its potential users. Existing management procedures are often difficult to change and project deadlines must be met, while application of the new technology may have legal implications and requires retraining of personnel.

The SEI has set a task for itself in education and in training. The purpose of the education program is to increase the number of qualified software engineers in the US. The purpose of the training program is to provide potential users of specific new techniques with hands-on experience in the use of the new tools or methods. The two tasks are clearly distinct. The first has a long-range goal that cannot be achieved by the SEI alone. The SEI can take the initiative and play a major role in the coordination of educational activities, but the work must be done in close collaboration with academic institutions. In order to make this happen, the SEI has started an academic affiliates program in which a dozen institutions work with the SEI on the development of educational material for software engineering. The status of the SEI's ongoing education project is discussed further on in this paper in the section describing the current projects. In contrast to the broad scope of the education program, the SEI's task in training is directed to teaching the use of specific techniques that are ready for transition. Training is provided for the methods, tools and environments that the SEI prepared for transition and that are on display in the SEI's software laboratory.

2.2. Topics of Interest

The three main topics of interest concern the existing tools and environments for software development, the issues of management and quality control and the development of a programming attitude that has the potential of saving a considerable amount of programmers' time. The three topics are not particularly intended to lead to specific projects that separately explore the issues involved, but are thought of as desirable aspects of projects undertaken by the SEI. The first two topics relate to existing technology that can be demonstrated by the SEI in prototype programming environments.

Reusability and automation are at this time the most promising techniques that, in combination with effective tools and programming environments, may substantially increase program productivity. Reusability is routinely practiced by sharing of operating system facilities and by the use of tools and libraries. However, most software systems contain large sections of code with very similar functionality that are written for each system from scratch. Reusability is often difficult to achieve because of poor documentation or programming language peculiarities or because of dependencies on the underlying operating system. These problems are difficult to avoid when programs are directly written in a programming language. Research into the topic of reusability tends to move in the direction of program descriptions at a level of abstraction that does not force the detailed bindings of a programming language. Tools, programming environments and automation play an important role in transforming such abstract descriptions into concrete programs that can be compiled or interpreted by existing language systems.

The SEI Plans

At this time it may be too early for the SEI to pursue reusability and automation because this topic has only recently received the attention of the research community. Instead, the SEI has scheduled its activities according to the relative maturity of the available technology. This criterion has led to the following sequence of phases planned for the initial five year period:

phase 1:   the use of the Ada language and associated tools;

phase 2:   techniques for integrating Ada and Ada tools with existingtools of other
           programming environments;

phase 3:   the transformation of programming environments to intelligent
           programming assistants;

phase 4:   the application of reusability and automation.

The phases will undoubtedly overlap in time. Each phase needs a period of preparation before specific projects can be started that lead to technology transition. We envisage that the focus of the SEI's activities will gradually shift from one phase to the next during the five year period. Phases will also not die out abruptly. It is for instance almost certain that the transition of Ada technology is still of great interest at the end of the five year period.

The SEI has started its activities with a series of workshops on the software factory concept which led to a broad discussion of the nature of software engineering. The first two workshops took place in March and April of 1985 with a group of approximately ten well-known scientists and engineers. The third workshop was held in October of 1985 with an attendance of approximately 50 people. The fourth and last workshop was held in February of 1986 with approximately 250 people attending.

One of the major results of the first two workshops was the decision to stand firmly behind the Ada language and to work on issues of programming-in-the-large related to Ada. The attendees of these workshops generally agreed that no major breakthrough is to be expected similar to the invention of the transistor for computer hardware that will all of a sudden solve the production problems associated with the development and maintenance of large systems. This opinion was shared by the attendees of the October and February workshops. It is generally believed that improvements must come from improved programming environment support combined with specific techniques that enables us to write reusable code and automate the program generation process.

Although there is general agreement among the experts in the field that a revolutionary breakthrough is not to be expected in the foreseeable future, there is no general agreement on the direction in which to push for an evolutionary improvement of the software production process. The various opinions were clearly stated by the members of a discussion panel at the February workshop. That particular panel consisted of two representatives of industry, one from the government and one from academia. The first panel member, representing the large software producers working on government contracts, took a strong position in favor of Ada based on the strong desire to make software portable. The representative of the government did not oppose the use of Ada, but was much more interested in the successive phases of the software production process, starting with requirement specifications and moving through design and functional specifications to decoding, testing and maintenance. The government representative expressed as his strongest desire the standardization of the various phases of the production process. The representative of the software industry took a strong position in favor of development tools and integrated programming environments for programming-in-the-large. The representative of the academic world took a position close to that of the representative of the software industry, but made a strong pitch in favor of Artificial Intelligence Technology, particularly for the rich interpretive environment of Lisp.

The best course is for the SEI to combine these ideas into a strategy that is based on the use of Ada, but emphasizes programming-in-the-large, and that moves the evolution of programming environments in the direction of the intelligent assistant in the style of Artificial Intelligence. The large software producers are right in stressing portability. The use of a standard Ada is a step in the right direction to achieve portability. Standardization is undoubtedly desirable, but may come too early for issues that have not stabilized. Standardization is possible for the Ada language and its compilers. It is, however, too early to standardize programming environments or the phases of the software production process. There are important developments taking place in the design of programming environments for Ada that each should be given a chance to mature. Regarding the Lisp environment, there is no reason why an Ada environment can not provide a similar functionality and a similar smooth user interface. The issue is not whether creating such an environment is possible for Ada, but whether someone will do the work. The Lisp environment took more than ten years to evolve!

4. Current Projects

The SEI has organized most of its activities as projects based on the areas and topics of interest and on the phases which determine the shifting emphasis over the years. In the eighteen months of its existence, the SEI has started eight projects on the following subjects:

1. the Software Factory Concept

2. the Showcase Environment (with the Ada Browser)

3. Software Licensing

4. Software Engineering Curriculum

5. Ada Applications

6. Ada Technology

7. Ada Environments

8. Intelligent Programmer Assistants.

The Software Factory concept was the central topic at the series of workshops discussed in the preceding section. The workshops have been very helpful for the SEI in shaping its five year plan and in determining its areas and topics of interest.

1.1. The Showcase Environment

The Showcase Environment and the Software Licensing project were started soon after the SEI was established. The Showcase project is a continuing effort to build up and maintain a software laboratory in which software development tools and environments are on display. The laboratory is used for demonstrations and for training. It has been used to show the merits of software development environments such as DSEE [8], the planning tools on MacIntosh, the object-oriented programming style of SmallTalk [9], the evolution of user interfaces as in Andrew [10], etc. A project particularly worth mentioning is the Ada Browser, which provides editing and tracing facilities for the visible parts of Ada packages. The designer of a software system written in Ada can look at his or her design in different views that show the interdependency of packages and can zoom in on packages to inspect details such as the types of sub-program parameters.

1.2. Software Licensing

The Legal Issues project is a first step in the area of the Nature of Technology Transition. The government is for instance faced with the fact that software it owns was written with tools it does not own. Questions arise as to how the government can assure that these tools will remain in existence for system maintenance. Serious problems exist with regard to intellectual property rights, software licensing, copyrights, etc. The first year of this project was concluded with a workshop held in April 1986 which was attended by many of the legal experts on software in the US. The results of the project are presented in a report that has been submitted to the SEI's sponsors.

1.3. Software Engineering Education

The Software Engineering Curriculum project is the main project in the Academic Affiliates program. Its initial purpose is the design of material that can be used in software engineering courses. In a later phase, the project will focus on programming environments for software engineering education. The team working on the course material consists of a small permanent SEI staff and a number of visiting faculty from various universities who spend a sabbatical semester or the summer at the SEI. The relationship between these visitors and the SEI will be extended over a number of years, which provides the opportunity for visitors to work on long-range plans.

The team working on the curriculum design came to the conclusion that specialization in software engineering fits best at the senior undergraduate and at the masters graduate level. The course material is designed for a masters degree in software engineering. The conclusion is based on the observation that software engineering must rest on a solid undergraduate education in computer science that contains components of discrete mathematics and electrical engineering. The design includes a description of the prerequisites for the software engineering curriculum.

The initial plans of the education team were discussed in detail with a group of approximately 25 experts at a workshop in April 1986. The attendees were unanimous in their opinion that an important part of a software engineering curriculum must consist of hands-on experience with real software systems. Small homework problems are of little interest because their solutions hardly ever extrapolate to large systems. Students should participate in an on-going project for an extended period of time during the summer or while taking the software engineering courses.

It was generally agreed that a software engineering curriculum should not be based on an enumeration of topics, but on lasting principles. The attendees distinguished between the general scientific principles of theory and experiment and the discipline-specific principles that characterize the activities in a field. A collection of the principles that are believed to distinguish software engineering from disciplines such as physics, mathematics and psychology are:

- the embedded character of software systems
- the discrete character of software
- the limited knowledge of the resources
- the decomposability of software problems.

Software engineering may share each of these principles taken separately with some other discipline, but the collection is what makes it clearly distinct from other disciplines.

Software systems usually perform a particular function in a larger organization that exists outside of the software world. This embedded character imposes evaluation criteria on software that depend on extraneous factors determined by the larger organization in which the embedded software system may play only a minor part. Software engineering is in this respect quite different from physics which explicitly states as one of its principles that its laws are universal. (There is in this respect more resemblance to biology if one considers the relationship between a particular animal and its habitat.)

Another distinction between software engineering and physics is the discrete character of software engineering problems in contrast to the continuous view physicists have of the natural world. The discrete character of software engineering brings the discipline closer to mathematics with which it has in common the basic concepts of formal models, enumeration, abstraction and inductive proof. Software engineering is, on the other hand, quite distinct from mathematics in that its objects do not have complete control of the resources they use. System behavior may depend on operating system facilites and hardware that was designed independently, or on the degree of concurrency which may vary from time to time. Such factors do not play a role in mathematical theories.

Decomposability might also be called the principle of divide and conquer. In this respect software engineering is again comparable to mathematics, but different from physics or psychology. Mathematicians and software engineers believe that they can decompose a problem, solve the parts separately and combine the solutions of the subproblems into a solution of the initial problem. Physicists and psychologists work under the hypothesis that the whole is larger than the sum of the parts which implies that there is always an aspect of the initial problem that has escaped consideration when one attempts to partition a problem of their domain.

The discussion on the principles of software engineering has given the education team a guide as to how to present various topics in the curriculum. Other sources of information that conributed to shaping their plans were the experience at Wang Institute and the perspective on future developments which largely coincides with the observations of the Software Factory workshops. A digest of all the material and ideas collected by the education team has led to a contract with a well-known publisher for a series of monographs on a variety of topics in software engineering.

1.4. Ada Applications

The Application project resulted from the October workshop on the Software Factory concept. A review of the state of Ada at that workshop resulted in a three part conclusion:

1. The initial difficulties with writing compilers for Ada have been overcome. More than a dozen Ada compilers have passed the validation test. Although Ada compilers now compile correctly, a lot of work is still needed in making the object code more efficient and in building optimizing compilers.

2. The Stoneman Report on the APSE Ada environment is based on outdated technology of the mid-seventies. It is too early to freeze the design or programming environments for Ada. There are several developments going on that should be tested on the commercial market.

3. There is a lack of good teaching material for Ada. There are several beginners texts, but most take a bottom up approach. The Ada reference manual is of good quality for the expert Ada compiler writer, but not suitable as an instruction manual. Teaching material is needed that takes a top down approach starting with the central modularity concept of Ada implemented by packages.

It seems that software producers have trouble introducing Ada, because the time can hardly be spared for teaching programmers the proper use of Ada. The lack of time for Ada instruction is primarily caused by the time commitments to government contracts that exclude all activities from the budget that do not directly contribute to the goal of that contract.

A working group at the October workshop discussed a plan to create a government contract specifically for the purpose of introducing Ada. The discussion resulted in an SEI project in collaboration with the STARS program (which is another component of the DoD's Software Initiative). The project involves the introduction of Ada in a number of industries that are interested in comparing the use of Ada with other languages they have used and are willing to experiment with Ada support tools or Ada environments. The industries will write a proposal for rewriting a system of more than 100,000 lines of code in Ada. Part of the task is to observe what kind of development tools are used to support Ada and how effective these have been. The task of the SEI consists of monitoring the projects, organizing meetings to let the participants "compare notes", and summarizing the results.

4.5. Ada Technology

The SEI tries to make use of the comparative advantage that it has because of being part of Carnegie Mellon University and because of the particular expertise of the people at the SEI. A field in which the SEI has particular expertise is in compiling techniques and automatic compiler generation. Since there is still room for improvement in the area of building Ada compilers, the SEI has started an Ada technology project that focuses on compiling techniques and tools for compiler construction.

The tools provided by this project center around the IDL description language that has been used to write the well-known DIANA description of Ada. The project is undertaken in collaboration with the University of North Carolina. The results will be made available to compiler writers and training in the use of the tools will be provided by the SEI.

4.6. Ada Environments

The Request for Proposal that came out in the spring of 1984 asked for an evaluation of the ALS and AIE Ada environments that had been contracted by the US Army and the US Airforce respectively. At the time the proposal was submitted in the summer of 1984, it became clear that the ALS system would still be in its early stages and that the AIE system would not be available at all. (The AIE Ada Compiler has finally been announced for September 1986 without an Ada Environment!) Since by this time several Ada compilers were reaching the market we proposed to explore the environments that these compilers were using instead of evaluating a particular environment. Final discussions resulted in the SEI's Ada Environment Evaluation project that has as its main goal to create a method for evaluating these vastly different environments.

The term "evaluation" in the project title is somewhat misleading, because no specific evaluation is contemplated, but a study is planned of how Ada environments can be evaluated [11]. The project has just concluded its first year with interesting results and plans to continue with a study of how Ada runtime and dynamic debugging environments could be evaluated. The Ada environments project has primarily focused on issues of programming-in-the-large. It did not want to repeat the elaborate validation process that Ada compilers have to go through to get official approval. The project also paid little attention to issues of code efficiency or optimization. The project looked at an Ada compiler as just one of the tools in a programming environment. If one looks at a compiler that way, issues of importance are things such as the quality of the error messages, the compilation speed, the linking procedures, etc. The general issues of primary interest in the project were those of system version control, configuration management, access control and project management.

It was clear from the start of this project that programming environments are hard to compare. Instead of basing the study on a comparison, we categorized the issues into four areas that respectively concern

- functionality
- user interface
- context utilization
- performance.

One can easily generate a list of functions that an environment must provide in order to handle source versions, documents and object code. One can also specify in general terms what kind of facilities must be available for system configuration management and for system construction. Environments may offer these functions in different forms or in different sets, but expert programmers will soon find out whether or not the necessary facilities are present.

The situation is slightly different with the user interface. There are some general characteristics that every user interface must have, but, in contrast to the functionality, user interfaces tend to have specific characteristics that strongly depend on the type of input-output device used. Among the general characteristics are properties such as "the response to an editing command must be less than a second", "the interface must distinguish between an expert user and a novice" or "for long operations, the interface must show at regular intervals that the system is still alive". The specific characteristics, however, depend on factors such as having a bitmap display or character terminal, whether mouse control is included, etc. Although the facilities offfered by these devices differ too much to make a useful comparison, it is not hard to list for each system separately some properties one expects a good user interface to have. We did that in the preliminary report of the project that was submitted to the SEI's sponsors [12].

Another criterion that determines the quality of a programming environment is the degree to which the available hardware and software resources are utilized. A good programming en-

vironment will make use of the potential execution speed of a large machine, will not do super-
fluous paging when sufficient memory space is available, will make use of existing software
and existing network facilities, etc. For each host machine and each host operating system, one
must determine which specific facilities are provided and how well the programming environment makes use of these facilities.

Performance has already been mentioned in passing in the paragraph on user interfaces. It is of
course a matter of concern for every part and every aspect of a programming environment. Our
study shows that at this stage of the development of programming environments special attention should be paid to the performance of linking procedures. Some compilers give the impression of being fast because programs are compiled in small pieces, but the advantage is lost
when it comes to linking modules into systems which for some language processors is not much
faster than if all modules were compiled together.

In the study of Ada programming environments, we found that designers have introduced
three different models which are represented by the Rational machine, the DEC-Ada system
and the ALS system. The philosophy of the Rational approach is that everything is Ada and
hardware is specifically designed to make this approach feasible. Ada is not only used as the
language for writing programs, it is also used for programming-in-the-large. The interactive
command language is for Ada, the user can write Ada command procedures, version control is
done with Ada libraries. The only other thing besides Ada is the concept of text which is
needed for documentation.

The opposite approach was taken by the designers of DEC-Ada. Here the philosophy was to
integrate Ada facilities with an existing environment which already provides extensive support
for programming-in-the-large and also offers the benefit of extensive program libraries. This
approach makes it possible for the user to live in a multi-language environment in which subprograms written in various languages such as Ada, Pascal and FORTRAN can be used in each
of the language systems. The user of DEC-Ada does not work in the uniform Ada world as the
user of the Rational machine, but has the benefit of using a rich environment of mature
software.

The ALS approach is in the tradition of operating system design. It is based on the view that an
Ada environment should be an interface between an operating system and its users in the style
of a Database Management System. In this approach, each user command can be treated as a
transaction that is controlled by the environment which performs all the necessary actions that
have to do with version control, access control, etc. The idea behind this approach is reflected
in the name ALS which stands for Ada Language System. The ALS differs most from the other
two in the automatic checks it performs and in including management procedures as part of the
standard user interface. A consequence of this approach is that ALS forms a layer on top of an
existing system that hides the underlying system from the user.

.7. Intelligent Program Assistants

Ve expect that programming environments of the future will show more intelligent behavior
han today's rigid environments. The term "intelligent" is used here in the behavioral sense of
what one expects an intelligent being to do". We expect that the development of the field will
ead to a type of environment that interacts with its users more in the style of an assistant than
f a toolbox. One of the main differences between an assistant and a toolbox is that the assistant
vill need little instruction and will apply his own judgement automatically as to what should
e done, possibly without involvement of the boss. Another main difference is that the assistant
as a large tolerance in understanding his boss. He is able to interpret commands depending
n properties of the comminication language and on factors such as the context and the history
f the interaction between the assistant and the boss.

he technology is emerging that makes this kind of environment possible. It is, however, not
et ready for transition into routine practice. We do believe that this technology has the atten-
ion of the research community and that the SEI should prepare for technology transition in this
rea during the last phase of the current five year plan. An important part of the preparation is
lanned as an SEI project in which we explore what technology already exists that may lead to
he desired goal. It is not surprising that the SEI looked in the first place at AI technology. At
his time the project studies in particular the various Lisp environments and the extensive assis-
ance provided by these environments for program development. The intermediate results are
nade available in the SEI's software laboratory.

he strength of a Lisp environment is not so much in the language, but in the rich collection of
acilities provided by the environment. We believe that the functionality of a Lisp environment
ould also be provided by an Ada environment. However, Ada is ten years behind the Lisp
levelopment. It will take a major effort to create an Ada environment that is as rich as a Lisp
nvironment. We believe that such an effort should be given high priority because
rogrammers' productivity will not increase enough without it.

ne feature of the Lisp environment does not easily carry over to an Ada environment: the
nterpretive nature of Lisp. The great advantage of the interpretive approach is that one does
ot need a dynamic debugger that operates on the object code. In a Lisp environment one can
mmediately test a fuction in the same environment in which it is written. Efficient object code
s produced at a later stage by a Lisp compiler when debugging with the interpreter has been
ompleted.

. Summary

he Software Engineering Institute was started as the result of the DoD's Software Initiative.
he primary purpose of the SEI is the transition of promising software development technology
rom the research environment to routine practice. The SEI has put together a five year plan
nd has scheduled successive phases in which it will introduce increasingly advanced tech-

nology. The first phase involves the Ada language and the tools that are needed for construct ing large software systems.

The SEI discussed its plans extensively with the leaders in the field of software engineering and with the software producers and users at large. As a result of these discussions, the SEI staf found its viewpoint confirmed that it should not adopt a particular software developmen methodology. Instead, the SEI will show how improvement of the production process can be achieved by replacing the traditional labor-intensive approach to managing a software projec by a technology-intensive approach. The SEI demonstrates the technology that is ready for transition in its software laboratory in which it also offers opportunity for training.

The SEI has organized most of its work in project form. It has started eight projects that dea with legal issues, with existing technology and with future developments. Although basi research and education are not the primary concerns of the SEI, they constitute an integral par of the program. In education, the focus is on a software engineering curriculum at the graduate level. The education project is a joint effort of the SEI and a number of participating academic institutions.

The first results of the work at the SEI are becoming available. The Software Licensing projec has submitted a report that has served as the basis for a public debate on software ownership and licensing arrangements. The Ada Environment project has produced a report in which i publishes the results of the experiments it conducted for the validation of its method for the evaluation of Ada environments. The Showcase project has built up the SEI's software laboratory in which several programming environments and development tools are on display.

In conclusion, the SEI has started a number of activities that directly serve its objective of tech nology transition. The five year plan specifies definite goals that are being achieved by a num ber of projects that initially focus on technology around the Ada language. The SEI will be successful if it can maintain and extend its contacts with the potential users of the advanced software development technology it is able to demonstrate.

References

[1]   Parnas, D. L.
      "On the Criteria to be Used in Decomposing Systems into Modules"
      Comm. ACM, Vol.15, No.12 (Dec 1972)

[2]   Martin, E. W.
      "Strategy for a Department of Defense Software Initiative"
      Computer, Vol.16, No.3 (Mar 1983)

[3]   "Reference Manual for the Ada Programming Language"
      US Department of Defense (Jan 1983)

[4]   DeRemer, F. and H. Kron
      "Programming-in-the-Large Versus Programming-in-the-Small"
      ACM SIGPLAN Notices (Jun 1975)

[5]   Kaiser, G. E. and A. N. Habermann
      "An Environment for System Version Control"
      Proc. IEEE Spring CompCon, San Francisco (Feb 1983)

[6]   Barbacci, M. R., A. N. Habermann and M. Shaw
      "The Software Engineering Institute: Bridging Practice and Potential"
      IEEE Software, Vol.2, No.6 (Nov 1985)

[7]   Boehm, B.
      "Software Engineering Economy"
      Prentice Hall, Englewood Cliffs (Jun 1981)

[8]   Leblang, D. B. and R. P. Chase, Jr.
      "Computer-Aided Software Engineering in a Distributed Workstation Environment"
      Proc. ACM SIGSOFT/SIGPLAN Symposium on Practical Software Development
        Environments
      Pittsburgh, Pa. (Apr 1984)

[9]   Goldberg, A. and D. Robson
      "Smalltalk-80: The Language and its Implementation"
      Addison & Wesley, Reading, Mass. (Mar 1983)

[10] Morris, J., J. Howard and F. Hansen
      "The Andrew Windowing System"
      Technical Report, Carnegie Mellon University (Jan 1986)

[11] Weiderman, N. H., A. N. Habermann, M. W. Borger and M. H. Klein
      "A Method for Evaluating Environments"
      Second ACM SIGSOFT/SIGPLAN Symposium on Practical Software
      Development Environments, Portland, Oregon (Jan 87)

[12] Weiderman, N. H., A. N. Habermann, M. W. Borger and M. H. Klein
      "Preliminary Evaluation of the Ada Language System"
      Technical Report SEI-86-MR-6, Carnegie Mellon University (Apr 1986)

Task Sequencing Language
for
Specifying Distributed Ada[1] Systems

D.C. Luckham
D.P. Helmbold
S. Meldal
D.L. Bryan
M.A. Haberler

Program Analysis and Verification Group
Computer Systems Laboratory
Stanford University
Stanford, California 94305

This research was supported by the Advanced Research Projects Agency, Department of Defense, under Contract N00039-84-C-0211 and by the Air Force Office of Scientific Research under Grant AFOSR 83-0255.

Abstract

TSL-1 is a language for specifying sequences of tasking events occuring in the execution of distributed Ada programs. Such specifications are intended primarily for testing and debugging of Ada tasking programs, although they can also be applied in designing programs. TSL-1 specifications are included in an Ada program as formal comments. They express constraints to be satisfied by the sequences of actual tasking events. An Ada program is consistent with its TSL-1 specifications if its runtime behavior always satisfies them. This paper presents an overview of TSL-1. The features of the language are described informally, and examples illustrating the use of TSL-1, both for debugging and for specification of tasking programs, are given. A definition of robust TSL-1 specifications that takes into account uncertainty in runtime observation of behavior of distributed systems is given. A runtime monitor for checking

---

[1] Ada is a registered trademark of the U.S. Government (Ada Joint Program Office).

consistency of an Ada program with TSL-1 specifications has been implemented. In the future, constructs for defining abstract tasks will be added to TSL-1, forming a new language, TSL-2, for the specification of distributed systems prior to their implementation in any particular programming language.

1. Introduction

Previously, we have studied techniques for automatically detecting tasking errors in Ada programs by runtime monitoring [7]. We demonstrated that classical kinds of errors such as deadlock and global blocking can be detected by our methods when the errors result from misuse of task rendezous. We also concluded that specifications expressing the intended order and synchronization of interactions between tasks could be monitored at runtime by similar techniques. Such specifications are necessary in order to detect task communication errors when tasks communicate by the most general means allowed in Ada, e.g., by sharing global variables, or other ways than the "preferred" Ada rendezvous. These specifications would be expressed in a suitable language, and supplied by the programmer with the program. The current version of TSL-1 described here is an outgrowth of our earlier work [8].

TSL-1 is a language designed to meet the following goals concerning power of expression, ease of use, and implementability:

1. any communication between tasks in an Ada program should be specifiable in TSL-1, whether communication takes place by Ada rendezvous or by any other means,
2. TSL-1 should be powerful enough to express general patterns of task communication,
3. it should be easy to specify simple, often used, kinds of task communication in TSL-1,
4. it should be within the current state-of-the-art to provide practical algorithms for checking consistency at runtime, or to provide rules for proving consistency between the behavior of an Ada program and TSL-1 specifications.

A consequence of the first goal is that TSL-1 must be related to, and compatible with, the Ada programming language. A key question is how close this relationship has to be. For example, TSL-1 obeys similar scope and visibility rules, and naming conventions. It is an addition to Ada. A consequence of the third and fourth goals is that TSL-1 sacrifices some power of expression for simplicity and implementability. For example TSL-1 is capable of expressing only bounded temporal dependencies — i.e., those that must occur before a specific event such as termination of a task.

TSL-1 provides three kinds of constructs for specifying distributed systems. The first kind are *actions*. Actions specify events in the system which are deemed significant and are to be included in the history of events whenever they occur during a computation. The second kind are *specifications* of patterns of events that should (or should not) occur in the history of events. The third kind are definitions of *properties* of entities in the system, where the properties depend on the history of events and may change their values as new events happen.

TSL-1 constructs are placed in the text of an Ada program — called the *underlying* Ada program. They appear as Ada comments and therefore have no effect when processed by standard Ada tools.

The meaning of TSL-1 constructs is described in terms of runtime checking. They are recognized by a TSL-1 compiler, which transforms them into Ada data structures and calls to a TSL-1 runtime monitor. At runtime the specifications (in their compiled data structure form) are passed to a TSL-1 monitor as the scopes in which they occur are elaborated and executed. As the underlying program executes, events are generated. The resulting history of events is ordered linearly, and will henceforth be called the *stream of events* (or *event stream*). The *stream of events* that occurs during the execution of the underlying Ada program is monitored for consistency with the TSL-1 specifications. The description of *monitoring* in terms of a process called *matching* defines the meaning of TSL-1 specifications and properties.

A fundamental concept in TSL-1 specifications is *order* of events. However, the order in which events from a computation of an underlying distributed Ada program appear in the event stream, as viewed by the TSL-1 monitor, is not always significant. That one event preceeds another in the stream of events passing by an observer may often tell that observer nothing. This uncertainty is a consequence of Ada semantics and the freedom accorded to implementations of Ada. It is unavoidable in any practical method of observing a distributed computation which permits many processors to operate independently. Fortunately, certain pairs of events do occur in the underlying Ada computation in a definite order, and it is assumed that their order in the event stream is the same. These are called *connected* events. The assumption requires the event stream implementation to be such that connected events must occur in the same order in all event streams that can arise from the same computation. So called *robust* specifications may be constructed using connected events. Robust specifications cannot be satisfied by one event stream and violated by another event stream arising from the same computation.

At present, a TSL-1 runtime monitor has been implemented using techniques generalizing those presented in [5, 6]. A compiler that preprocesses TSL-1 into an Ada form acceptable by the monitor has also been implemented. We refer to the TSL-1 compiler and monitor together as the TSL-1 system. This system is intended primarily for use in testing and debugging tasking behavior. It should enable the programmer to carry out these activities using the same concepts and language as were used to specify and build the program.

It also turns out that TSL-1 can be retargeted to specify concurrent behavior of systems in languages other than Ada simply be redefining the set of basic events (Section 3). Thus TSL-1 is envisaged as the basic part of a new language, TSL-2, for specifying behavior of distributed systems independently of the target programming language. Design constructs in TSL-2 will provide an ability to define hierarchies of abstract processes and actions, and to specify temporal relationships between sets of actions. TSL-1 will continue to provide the basic facilities for specifying behavior at any given level in the TSL-2 design. TSL-2 will allow a simulation of a system at a given level of abstraction to be monitored for consistency with behavioral specifications referring to the actions defined at that level. An Ada tasking program is viewed as representing a distributed system at a particular level of detail in a TSL-2 design. Rendezvous constructs, for example, would be represented as actions. These actions would have the connection properties derivable from the semantics of Ada as described in Chapter 9 of the Ada reference manual [1].

This paper gives an overview of TSL-1. A restricted form of TSL-1 is described first, and then the description of the general form is based on that. Section 3 is the overview, and Section 5 gives examples. Methods of constructing robust TSL-1 specifications using connected events are discussed in Section 4. Work published by other researchers which either predates TSL-1 and influenced our thinking, or is concurrent with TSL-1 and presents similar ideas, is discussed in Section 6.

A separate report is in preparation which will give a detailed description of the matching and runtime checking concepts of TSL-1 in terms of abstract computations on directed labelled graphs with tokens. This will serve to fill in gaps in the informal description of the meaning of various constructs given here. It will also indicate how runtime checking can be implemented. Other reports on the Ada design of the TSL-1 runtime monitor, on the relationship of TSL-1 to Temporal Logic, and on the applications of TSL-1 to Ada are in preparation.

2. Illustrative Problems

This introduction to TSL-1 will draw on two simple Ada tasking programs to give examples illustrating various points. The programs are described informally below. The reader will be able to follow the examples from the descriptions. In

2.1 Automated Gas Station

The Ada program simulates an automated gas station [7]. The gas station contains tasks representing the station operator, the customers, and the pumps. A customer arrives at the station, prepays the operator, pumps gas, receives change from the operator, and leaves. The operator accepts prepayments, activates pumps, computes charges and gives customers their change. The pumps are used by the customers, and report the amount pumped to the operator.

2.2 Reliable Soldiers Problem

The Ada program represents a General and a number of soldiers as tasks. The General sends an order to each soldier, either "advance" or "retreat". Each soldier subsequently sends a message to each of his comrades, which is supposedly the order he received, but he might lie sometimes. When each soldier has received a copy of all of his comrades orders, he takes an action (we do not care how the action is decided upon). A soldier is *reliable* if he sends the order that he received from the General to everyone [11]. The problem is to specify that all reliable soldiers take the same action.

3. Overview of TSL-1

The semantics of TSL-1 are based on a model of observing the computation of an Ada program whereby events occuring during the computation are organized into an ordered linear sequence. An event may signify something in the computation of a single thread of control, or it may signify something involving more than one thread of control, such as the start of a rendezvous between two tasks. This sequence of events is called the *event stream*.

The threads of control may be executing on many processors. It is assumed that the events occuring in a single thread of control appear in the stream in the order in which they were executed. When two events occur in separate threads of control, only a weak assumption is made regarding their order in the event stream. Namely, only those pairs of events that are *connected* — to be defined later — are assumed to occur in the event stream in the order in which they occur in the computation of the underlying program. For example, if A and B are tasks, the events "A **calls** B" and "B **accepts** A" associated with a particular rendezvous are always connected events and must occur in this order both in the Ada computation and in the stream. On the other hand, "A **calls** B at E" and "B **accepting** E" need not be connected events, depending on the Ada program executing them, and their order in the event stream may or may not be significant. Under the assumption that only pairs of connected events are required to have the same order in the event stream as their order of execution in the Ada program, the event stream can be implemented in a practical and efficient manner. We discuss connected events in Section 4.

TSL-1 provides three simple capabilities: (1) declaration of *actions*, and *perform* statements which refer to actions and have the effect of inserting corresponding events in the event stream, (2) declaration of constraints on the event stream, called *specifications*, and (3) definition of simple kinds of *properties* associated with entities in the underlying program, the values of which depend on the event stream.

TSL-1 also provides predefined actions and properties relating to the basic tasking constructs of Ada. The predefined actions include accepting an entry, issuing an entry call, and activating a task. For example, "the OPERATOR task is **accepting** entries PRE_PAY and CHARGE" is an event generated when the OPERATOR task executes a select statement with open accept alternatives for those entries. The predefined properties include whether or not a task is running, blocked, or terminated, if an entry of a task is open, and so on. For example, "BLOCKED(OPERATOR)" is a boolean-valued property that is TRUE at certain points in the event stream — e.g., if the most recent event in the stream signifies that the OPERATOR has executed an accept statement and previous events signify that no task has both executed a call to the corresponding entry and is waiting for the call to terminate; it will remain true until new events signifying calls to the operator appear in the event stream.

TSL-1 declarations and statements appear as comments in the *underlying* Ada program. These comments are prefaced by the TSL-1 comment symbols, "--+" indicating that they are to be processed by the TSL-1 system. A TSL-1 declaration is placed in a declarative part of the Ada program. TSL-1 declarations are action declarations, specifications (including macros), property declarations, and property bodies (specifying how to compute property values). A TSL-1 *perform* statement for an action, is placed in a sequence of statements of the Ada program within the scope of the action declaration. It is compiled into Ada text which inserts a structure representing an event into the event stream whenever the Ada text is executed.

The meanings of TSL-1 constructs are described in terms of the event stream. Consider, for example, the case of TSL-1 *specifications*. First of all, the event stream, representing activity in the underlying program, is generated by Ada statements inserted into the underlying program by the TSL-1 compiler. A TSL-1 specification contains a pattern that can match many different sequences of events, and it expresses whether its pattern is required or forbidden to occur in the event stream of the program. That is, informally, a TSL-1 specification has the semantics: "a subsequence of the underlying event stream between such and such events must or must not match the pattern". Specifications usually cannot be checked immediately like an assertion. Instead, they must be checked against a segment of the event stream that is generated over a period of time during an Ada computation, − we call this *monitoring*.

A TSL-1 specification is *elaborated* when the declarative region containing the specification is elaborated. If the declarative region is elaborated more than once, a new instance of the specification is passed to the monitor by each elaborating thread of control. Elaboration involves making the specification ready for monitoring against the event stream. A copy of an elaborated specification may be *activated* when an appropriate subsequence of events occurs in the event stream. It will remain active until events occur in the event stream which imply either that it is *satisfied* or that it is *violated*. When an elaborating thread of control leaves a declarative region, all TSL-1 specifications declared in that region which were elaborated by that thread become inactive. Any currently active copies of these specifications are treated as if their terminating conditions had occurred.

Different kinds of error situations may arise in a TSL-1 computation. For example, a specification may be violated, indicating an inconsistency with the underlying Ada program. Other kinds of errors such as type mismatches may also occur. No specific action is required in error situations. Implementations of TSL-1 are free to take any action, including aborting the TSL-1 monitor and issuing error messages. Such actions will often be diagnostic in nature and will not disturb the computation of the underlying Ada program.

Events are expressions in TSL-1 and also components of the stream of events representing activity in the underlying program. We will use the term *event* to refer to events in TSL-1 specifications, and *stream event* to refer to an event in the underlying event stream.

Reserved words in TSL-1 are written in lowercase boldface. Names of TSL-1 predefined types and properties are in uppercase. Identifiers in TSL-1 may be written either all capitalized or only the first letters capitalized.

3.1 Types, Expressions, and Basic Events

There are no type declarations in TSL-1. The standard types of TSL-1 are (1) the TSL-1 types TASK, ENTRY, TASK_TYPE, ACTION, and (2) the Ada types INTEGER, BOOLEAN, and the other Ada discrete types. These types define domains of values and operations on those values. Values of type TASK are the task identifiers (abbreviation: *ids*) uniquely associated by TSL-1 with each activated thread of control during a computation. (The domain of values of task ids depends on the TSL-1 implementation.) The names of task objects in the underlying Ada program are constants of the TSL-1 type, TASK, so a task name may be used in TSL-1 expressions to designate its task id value. The constant, **main** always refers to the main program and **self** always denotes the task id of the executing task. The sets of values of the types ENTRY, TASK_TYPE, and ACTION are unique *ids* associated with names of task entries, and task types in the underlying Ada program, and with TSL-1 actions respectively. Entry names, task type names, and action names may be used as literals in a TSL-1 specification whenever they are visible in the Ada program; they must be fully expanded according to Ada rules, if necessary, to avoid ambiguity. The only operations on the TASK, ENTRY, TASK_TYPE, and ACTION types are the equality and inequality operations, and assignment. In addition, there is a literal **null** in each of these TSL-1 types that denotes a null value.

The types INTEGER, BOOLEAN, and the discrete types are the standard Ada types — i.e., in TSL-1 expressions the standard Ada operations associated with those types are allowed. (Note that the syntax summary lists only a subset of the Ada operations on these types; this reflects a restriction in the present TSL-1 monitor implementation.)

A task type declared in the Ada program defines a subtype of the TASK type in TSL-1. This subtype has the same name in TSL-1 as the Ada task type, and consists of those TSL-1 task ids associated with objects of the Ada task type.

TSL-1 contains two kinds of expressions, *primary expressions* denoting values, and *events*. Primary expressions are composed according to well known composition rules from literals, TSL-1 variables (called placeholders, described later), TSL-1 operators (a subset of Ada operators), and TSL-1 properties (which obey the same composition rules in expressions as Ada functions). Events in TSL-1 are composed out of one or more *basic events*.

A *basic event* denotes the performance of an *action* in the underlying program. There are six predefined classes of basic events (see syntax summary). The six predefined event classes denote Ada tasking actions. Each predefined class of event implies a particular TSL-1 type for each of its constituent expressions; these types are indicated in the syntax definition of an event class by the italic prefix of the names. For example, the start of rendezvous is a predefined event class. The TSL-1 syntax,

start_rendezvous ::= *task*_name **accepts** *task*_name **at** *entry*_name

indicates the types of the three constituent names. An example of a start rendezvous event is

OPERATOR **accepts** CUSTOMERS(5) **at** PRE_PAY

which specifies an event in which the operator starts a pre-pay rendezvous with the fifth customer task in the array of customers.

It is assumed that type mismatches never can occur in the event stream. Type mismatches may occur during a TSL-1 computation as described later, either in the matching process or in the evaluation of expressions. Generally such mismatches will be detectable only at runtime. Whenever a TSL-1 type mismatch occurs, the TSL-1 computation is erroneous. The handling of erroneous situations is left open to the implementation of TSL.

Each kind of basic event is said to be *generated* by one of the tasks that appear in the event. For example a **calls** event,

    T1 **calls** T2 **at** E

is generated by T1 (see appendix).

3.2 User-Defined Events

Execution of tasking statements in the underlying Ada program results in corresponding predefined TSL-1 events being entered into the event stream. These predefined events often do not include all of the significant events in a computation that are needed to specify the correct behavior. This is particularly true, for example, when tasks can communicate by means of shared data. TSL-1 therefore provides a facility allowing declaration of other events that should be entered into the event stream.

New classes of basic events are declared by *action declarations*. An action declaration may be placed in the position of either an Ada entry declaration in a task, or an Ada basic declaration. For example, suppose we wish to define as an event, the activity whereby the General sends an order to a soldier. An action declaration,

    action Order(M : Order_Type; S : TASK);

can be made either as a basic declaration in the Ada program, or else as an action of the General task; in the latter case, it is placed in the position of a task entry declaration. Here, Order_Type is a discrete type in the Ada program, and TASK is the TSL-1 standard task identifier type. The task type SOLDIER in the underlying program can also be used in the action declaration if it is visible at that point:

    action Order(M : Order_Type; S : SOLDIER).

SOLDIER defines a subtype of the TASK type in TSL-1.

This action declaration declares a new class of basic events of the form,

    T **performs** Order(M1, S1)

where T is the task id of the thread of control performing the action, and the types of constituent expressions in the event are those declared in the parameter list of the corresponding *action* declaration. In this example, M1 is of type Order_Type, and S1 is a task id of a task of type, SOLDIER (i.e., a member of the subtype of TASK defined by the type, SOLDIER). T is said to *generate* the performs event.

TSL-1 provides *perform statements* corresponding to user-declared actions. These are statements that may be placed in the underlying Ada program to indicate when, in the course of a computation, that action is actually performed. If the action is in the position of a basic declaration, then corresponding perform statements may be placed in the position of an Ada statement within the scope of the action declaration. If the action declaration is in the position of an entry in a task declaration, corresponding perform statements may be placed in the body of the task, but only where Ada **accept** statements for such an entry could be placed. This restriction ensures that an action in a task specification can only be performed by the task itself, and not by subordinate tasks.

Perform statements are executed when an Ada statement in the same position would be executed, and result in a stream event being placed in the stream. The parameters of a perform statement are treated as if they were mode **in** parameters of an Ada subprogram call. Therefore, the user, having declared an action, can place perform statements appropriately to define the meaning of that action in relation to other events taking place in the Ada program.

Example: An action declaration and perform statement.

```
    task General is
--+     action Order(X : Move; S : Task);
    end;

    task body General is
        M : Move := Decide(Advance);
    begin
        for I in 1 .. Max_Soldiers loop
--+         perform Order(M, Soldiers(I));
            Mailboxes(I) := M;
        end loop;
    end General;
```

Commentary

In this example the perform statement indicates that the action is performed when the General places an order in the soldier's mailbox. An event,

```
        General performs Order(M, Soldiers(I))
```

will enter the event stream when the assignment statement is executed. In more complex simulations of the Byzantine General's problem the Ada activity denoted by the action might involve a sequence of events such as the activation and use of messenger tasks to communicate the order to a soldier. In such cases, it may be necessary to declare other actions such as `Start_Order` and `Complete_Order` to correctly represent the relationship between this event and other events in the program.

Note that the predefined basic events are equivalent to user-defined actions when corresponding perform statements are placed in each program in appropriate positions in relationship to Ada tasking statements.

3.3 Matching

In TSL-1 *matching* is the fundamental operation on events, analagous to evaluation of expressions. A TSL-1 event *matches* an event from the underlying stream if the stream event is of the same class and has the same constituent values. The event,

OPERATOR **accepts** CUSTOMERS(5) **at** PRE_PAY

would *match* all of the start rendezvous stream events where the OPERATOR has accepted CUSTOMERS(5) at entry PRE_PAY. (Note that an instance of a TSL-1 event may occur in the stream many times as distinct stream events.)

We also use the concept of an event matching an interval. An *interval* of the event stream is a contiguous section of the event stream containing one or more stream events. A TSL-1 event matches an interval of the event stream if it matches a stream event in the interval.

3.4 Compound Events

There are four classes of compound events: *one_of*, *all_of*, *sequence*, and *iterative* events. They are constructed from finite sets of two or more events connected together by **or, and,** =>, and ↑ respectively. The constituent events of a compound event may be compound events.

A *one_of* compound event matches an interval of the event stream if any of its constituent events matches the interval. An *all_of* compound event matches an interval of the event stream if there is a one − one association between its constituent events and disjoint subintervals of the interval such that each constituent event matches the subinterval associated with it. A *sequence* of N events matches an interval of the event stream if the interval can be partitioned into N disjoint subintervals such that each event matches the corresponding subinterval and the correspondence preserves order (i.e. the first event in the sequence matches the first subinterval, etc.). An iterative event, A↑N, matches an interval if A matches N disjoint subintervals of the interval.

Note that when matching compound events: *(1)* non-matching stream events may occur between matching ones, and *(2)* constituent events must be matched by non-overlapping intervals. Also, an iterative event matches whenever its constituent event matches the specified number of times; if the integer exponent is not positive, the event matches every interval.

Examples of compound events and matching.

A one_of event:
    OPERATOR **accepting** PRE_PAY **or**
        OPERATOR **calls** CUSTOMERS(1) **at** CHANGE

An all_of event:
    OPERATOR **calls** CUSTOMERS(1) **at** CHANGE **and**
        CUSTOMERS(1) **accepting** CHANGE

A sequence event:
      CUSTOMERS(1) **calls** OPERATOR **at** PRE_PAY   =>
      CUSTOMERS(1) **calls** PUMPS(1) **at** START_PUMPING =>
      CUSTOMERS(2) **calls** PUMPS(1) **at** START_PUMPING

An iterative event:
    (OPERATOR **accepting** PRE_PAY)↑5

Matching compound events:
    if the event stream is: ..., A, C, B, D, A, E, B, ...
        the TSL-1 event, **A and** B *matches the subintervals:*
          [A, C, B], [B, D, A], [A, E, B]
        the TSL-1 event, A => B *matches subintervals:*
          [A, C, B], [A, E, B]
        the TSL-1 event: **(A and D)** => **(A and B)** *matches interval:*
          [A, C, B, D, A, E, B].
        the TSL-1 event, A↑2 *matches the subinterval:*
          [A, C, B, D, A].
In all cases the subintervals are the minimal matching intervals; intervals containing them also match.

The definition of matching implies certain equivalences between compound events. For example,
    (A **and** B)
matches an interval I if and only if (A => B) **or** (B => A) matches I.

3.5 Specifications

A TSL-1 *specification* consists of three events in the following order: an *activating event* (or *activator*), a *specified event* (or *body*), and a *terminating event* (or *terminator*). The activator begins with the keyword, **when**, and the terminator begins with one of the keywords, **before** or **until**. The specified event may be *negated*, in which case it is preceeded by the negation, **not**. The activating and terminating events cannot be negated. A specification is either *negative* or *positive* depending whether the specified event is negated or not.

It should be noted that all three component events in a specification may be compound events, although frequently the activator and terminator will be only basic events.

The semantics of a positive specification may be described informally as follows:

> Whenever the activating event matches an interval of the stream, and the last event of the interval is a matching event, then the specification is *activated* at that event. For that activation, both the specified event and the terminator simultaneously begin matching intervals of the event stream, starting at the next event in the stream and continuing until one of them matches an interval. Whenever the specified or terminating events match, that activation of the specification is completed. If the specified event matches before the terminator, the specification is *satisfied*. If the terminator matches before the specified event, the specification is *violated*. If they both match the same (shortest) interval, the specification is *satisfied* if the terminator begins with **until** and is *violated* if it begins with **before**. Finally, if the scope of an active specification is exited, any activation is treated as if the terminating event had matched.

If a specification is negated, the meanings of satisfaction and violation are reversed.

Example: Activation, satisfaction and violation of positive specifications.

In the event stream interval: ..., A, C, B, D, A, E, B, ...
the following TSL-1 specification is activated at each of the A events; the first activation is satisfied:

>   **when** A
>   **then**   C => D
>   **before** B => A;

the following TSL-1 specification is activated at each of the B *events;*
 the first activation is violated:

>   when (A => B)
>   then   (C => D)
>   before (D => A) ;

The activator determines when the specification becomes active. A **before** terminator indicates that the specified event must match *before* the terminator, whereas an **until** terminator indicates that the specified event must match by the time the terminator matches.

Example: Specification of a gas pump's protocol.

>   when PUMP accepts OPERATOR at TURN_ON
>   then PUMP accepts CUSTOMER at START_PUMPING =>
>   PUMP accepts CUSTOMER at FINISH_PUMPING =>
>   PUMP calls OPERATOR at CHARGE
>   before PUMP accepts OPERATOR at TURN_ON;

Commentary

The specification is enabled whenever the declarative region of the Ada program in which it occurs, is elaborated. It is then activated whenever an event occurs in the stream signifying that the PUMP accepts a rendezvous with the OPERATOR at its TURN_ON entry. As a result of such an event, an activation of the specification is created. For each activation, the specification requires that the PUMP must execute the three actions in the sequence (body of the specification) before the PUMP accepts the OPERATOR at TURN_ON again (terminating event). Since a terminating stream event also leads to another activation, the pump must continually satisfy its protocol.

Example: Specification denying one race condition for a pump.

>   when OPERATOR accepts CUSTOMERS(1) at PRE_PAY
>   then not OPERATOR accepts CUSTOMERS(2) at PRE_PAY =>
>   PUMP accepts CUSTOMERS(2) at START_PUMPING
>   before PUMP accepts CUSTOMERS(1) at START_PUMPING;

Commentary:

This specification will be passed to the TSL-1 monitor whenever an Ada declaration in its position would be elaborated. After elaboration, it is activated each time the OPERATOR accepts a prepayment from the first customer task, CUSTOMERS(1). The body of the specification will match any subsequent stream interval where the OPERATOR accepts a prepayment from CUSTOMER(2), who then gets ahead to use the PUMP before **CUSTOMER(1)**. This describes a *race* between two customers for the pump. Since the specification is negated it will be violated every time this particular race condition occurs.

The termination event for this specification is **CUSTOMER(1)** starting the pump. If this happens *before* the sequence in the body matches, the specification is *satisfied* since it is negated. Notice that the body and terminator cannot complete their matching at the same stream event, so replacing **before** by **until** would not change the meaning of the specification. It may happen that after activation neither the body nor the terminator match. This will be the case, for example, if **CUSTOMERS(1)** never uses the pump, and a second customer does not prepay the operator. When the declarative region of the specification is exited, the terminator is assumed to match, leading to satisfaction in this example.

Positive specifications express *liveness* during the execution of their scope (i.e., that something should happen). Negated specifications express *safety* (i.e., that something should not happen). Therefore upon scope exit all undecided (i.e., neither satisfied nor violated) activations of a positive specification will be violated, and all undecided activations of a negated specification will satisfied. The following equivalences hold provided all activations are decided before scope exit:

> **when** E1 **then not** E2 **before** E3;
is equivalent to
> **when** E1 **then** E3 **until** E2;

> **when** E1 **then not** E2 **until** E3;
is equivalent to
> **when** E1 **then** E3 **before** E2;

3.6 Placeholders and General Matching

The examples given so far of events and specifications clearly lack generality. The basic events must in fact be identical with the stream events in order to match. What is most often required is an ability to specify a set of stream events that differ at their component values. For example, we may wish to specify a TSL-1 event that will match any stream event in which the **OPERATOR** accepts a customer at **PRE_PAY**.

Variable parts of a basic event, compound event, or specification are indicated by identifiers beginning with the question mark symbol, ?. They are called *placeholders* since they indicate a position for any *value* of their type. There is no explicit placeholder declaration. The type of a placeholder is determined by its position in an event according to the syntax for that class of event if it is predefined, or according to the formal parameter list of the action declaration if it is user-defined. There is also a special wild-card placeholder, **any**, for each TSL-1 type, which obeys different matching rules from normal placeholders.

Examples: Basic events with placeholders.

```
GENERAL performs ORDER(Move => ?M, Soldier => ?S);
OPERATOR releases ?C from PRE_PAY(PUMP_ID => ?P);
```

Commentary

Intuitively, the first event will match any event where the general orders any soldier, ?S, to do ?M (the named form of parameter bindings is used in the ORDER action). The second event will match any event where the operator completes a rendezvous at its PRE_PAY entry with any task, ?C, for any pump, ?P.

The meaning of basic and compound events containing placeholders is defined by a more general concept of matching, which in turn depends on the concept of instantiation.

Instantiation.

A TSL-1 event or specification may be *instantiated* if there is an association of a unique value with each placeholder and with each position of **any**. The instantiation is the event or specification that results when each placeholder is replaced by the associated value at all of its positions, and **any** is replaced at each of its positions by the value associated with that position.

An instantiation of an event or specification is said to be a *partial instantiation* if not all of the placeholders have an associated value.

Matching.

A TSL-1 event or specification *matches* a stream event or interval if there is an instantiation of it which matches that stream event or interval according to the previous matching rules.

For example, the TSL-1 event:
OPERATOR **accepts** ?T **at** ?E
matches all stream events where the OPERATOR starts a rendezvous with any calling task at any entry.

The instantiation rules require a placeholder to be associated with the same value at all positions, whereas **any** may have a different value at each position. Thus the stream interval

```
[OPERATOR accepts CUSTOMER at PRE_PAY,
     OPERATOR accepts PUMP at CHARGE]
```

does not match the compound event,

```
OPERATOR accepts ?T1 at ?E and OPERATOR accepts ?T2 at ?E
```

because ?E cannot be instantiated to different entry names at different positions. However, it does match both of the following compound events:

```
(1) OPERATOR accepts ?T1 at any and
        OPERATOR accepts ?T2 at any;
(2) OPERATOR accepts ?T1 at ?E1 and
        OPERATOR accepts ?T2 at ?E2;
```

The semantics of specifications with placeholders is defined from the semantics for the previously discussed cases when there are no placeholders.

1. A specification is activated when there is an instantiation of its activator which matches an interval of the stream. Activation takes place at a stream event which is both a matching event and the last event of a matching interval. Such an event is said to complete a matching of an activator.
2. This instantiation is applied to both the specified and terminating events. The resulting events are the constituent specified and terminating events for that activation.
3. The specified and terminating events from step 2 are both matched concurrently against subsequent intervals of the event stream. The smallest interval following the activating event which matches (an instance of) one or both of these events is found. At this point the satisfaction or violation of the specification is determined from the definition for specifications without placeholders.

Note that a stream event can complete the matching of more than one instance of an activator thus leading to more than one activation.

Specifications containing placeholders must obey one restriction to maintain the independence of the specified and terminator events:

If a placeholder occurs in *both* the specified and terminating events then it must always be bound when the activator is matched (consequently it must also occur in the activator).

The previous example of a specification may now be stated in a more general form.

Example: Specification denying all instances of a race condition.

```
when OPERATOR accepts ?C1 at PRE_PAY
    then not OPERATOR accepts ?C2 at PRE_PAY    =>
             ?P1 accepts ?C2 at START_PUMPING  =>
             ?P1 accepts ?C1 at START_PUMPING
    until ?P2 accepts ?C1 at START_PUMPING;
```

Commentary

This specification will be activated at each stream event in which the OPERATOR accepts a prepayment from any customer. The sequence in the body will then match any subsequent stream interval where the OPERATOR accepts a prepayment from a (later) customer who then gets ahead to use the pump to be used by the first customer. Since the specification is negated, it will be satisfied if the first customer gets to use any pump *before* the specified sequence matches. If the second customer uses a pump before the first customer, then a race is averted if the first customer uses a different pump. If however, the first customer uses the same pump, after the second customer, a race has occurred in which the first customer lost. In this case the body and terminator both match an interval ending at the same event — ?P1 and ?P2 represent the same pump. To exclude this case, **until** must be used. The negated specification expresses "the body must not match before the terminator nor simultaneously with it".

Example: Activation and matching of a specification.

Suppose an interval of the stream of events is:
```
OPERATOR accepts  CUSTOMERS(1) at   PRE_PAY,        (event 1)
OPERATOR releases CUSTOMERS(1) from PRE_PAY,        (event 2)
OPERATOR accepts  CUSTOMERS(2) at   PRE_PAY,        (event 3)
OPERATOR releases CUSTOMERS(2) from PRE_PAY,        (event 4)
PUMP(1)  accepts  CUSTOMERS(2) at   START_PUMPING,  (event 5)
PUMP(1)  releases CUSTOMERS(2) from START_PUMPING,  (event 6)
PUMP(2)  accepts  CUSTOMERS(1) at   START_PUMPING,  (event 7)
```

The activator in the previous example matches event 1 by associating CUSTOMERS(1) with ?C1.
The instances of the body and terminator for the resulting activation are:
```
body_1:        OPERATOR accepts ?C2 at PRE_PAY    =>
               ?P1 accepts ?C2 at START_PUMPING  =>
               ?P1 accepts CUSTOMERS(1) at START_PUMPING
terminator_1:  ?P2 accepts CUSTOMERS(1) at START_PUMPING
```
Event 3 matches the first event of body_1 by associating CUSTOMERS(2) with ?C2.

```
body_2:          OPERATOR accepts CUSTOMERS(2) at PRE_PAY    =>
                 ?P1 accepts CUSTOMERS(2) at START_PUMPING   =>
                 ?P1 accepts CUSTOMERS(1) at START_PUMPING
```
Event 5 matches the second event of body_2 by associating PUMP(1) with ?P1.

```
body_3:          OPERATOR accepts CUSTOMERS(2) at PRE_PAY      =>
                 PUMP(1) accepts CUSTOMERS(2) at START_PUMPING =>
                 PUMP(1) accepts CUSTOMERS(1) at START_PUMPING
```
Event 7 matches terminator_1 by associating PUMP(2) with ?P2.

Commentary

The example shows that the association of values to placeholders,
```
     ?C1 => CUSTOMERS(1), ?C2 => CUSTOMERS(2),
     ?P1 => PUMP(1), ?P2 => PUMP(2)
```
defines an instantiation of the specification which is satisfied by the
stream interval [event 1, ..., event 7]. The instance is activated
by event 1. The terminator of this instance matches before the body
(the specification is negated). Notice that event 3 will also lead to
another activation of the specification in which CUSTOMERS(2) is
assigned to ?C1. Also, the placeholder ?C1 appears in both the body
and terminator. It must therefore appear in the activator, and it is
bound to CUSTOMERS(1) at the first activation.

If instead **event 7** had been:
```
     PUMP(1) accepts CUSTOMERS(1) at START_PUMPING
```
the body would have matched interval [event 1, ..., event 7] and
violated the specification – customers 1 and 2 would have been racing
for the same pump.

It should be noted that activators in specifications will normally be basic
events. Compound events as activators are allowed but can lead to specifications
with counter intuitive meanings (see, e.g., example at end of Section 3.7).

Type mismatches may occur during the matching of a specification. For
example, an instantiation may associate a value with a placeholder that matches
the type required at one position of the placeholder but not at another position.
Such situations are errors in the TSL-1 computation, and may indicate an
inconsistency with the underlying Ada program.

3.7 Properties

Properties may be associated with entities in the underlying Ada program by
TSL-1 property declarations. The types to which the property applies are
specified in the declaration; all entities of those types then have the property.
Properties may be viewed as user-defined dynamic attributes, analogous to

standard attributes provided by Ada such as E'COUNT and T'CALLABLE. For example, we can declare a property of pumps at the Gas Station called In_Use.

Example: A property declaration.

```
property In_Use(PUMP) : BOOLEAN := FALSE;
```

The *values* of properties are functions of the event stream. These values are computed by *update* statements.

Update statements behave as concurrent processes which match specifications against intervals in the event stream, and update property values whenever the matching process is successful. An update statement contains an *activator* event, a *specified* event, and an assignment to a property. The assignment part is prefaced by the reserved word, **set**. Whenever the activating event matches, the update statement is activated. Subsequently, whenever the specified event matches, the assignment is executed, and that activation completes. The update may be re-activated by another matching of its activator.

The update statements for a property are included in a body associated with the property declaration. A property and its body may be declared separately in the same Ada declarative region. This is similar to Ada scope rules for subprogram declarations and bodies, but also allows a property declaration to be in a task specification and the property body to be placed in the task body. Mutually recursive properties are allowed but should be used with caution (see below).

Example: In_Use property declaration and body.

```
property In_Use(PUMP) : BOOLEAN := FALSE
is
     when ?X accepts any at START_PUMPING
        then set In_Use(?X) := TRUE;
     when ?X accepts any at FINISH_PUMPING
        then OPERATOR releases ?X from CHARGE
              set In_Use(?X) := FALSE;
end In_Use;
```

Commentary

The specification part declares a boolean valued property of every value of the subtype, PUMP, of the TSL-1 type TASK. It has an initial value, FALSE. The following body contains a sequence of *update* statements. After the property definition is elaborated, the update statements are all matched concurrently against the event stream. In this example, whenever the activator in the first update matches, the property for that value of the placeholder, ?X, is set to TRUE. Concurrently, whenever the activator in the second update matches, the specified event is activated; at a subsequent stream event that completes the matching of the specified event, the property for the value of ?X is set to FALSE. Consequently, the In_Use property for each pump changes its value when the pump accepts its START_PUMPING entry, and when it accepts FINISH_PUMPING and then finishes a rendezvous with the OPERATOR at CHARGE.

Example: A task's most recent action.

```
property Last_Done(TASK) : ACTION := null
is
     when ?T performs ?A
          then set Last_Done(?T) := ?A;
end Last_Done;
```

Commentary

Last_Done is a property of tasks yielding values of the ACTION type. Initially the property has the **null** value. Each time a task performs an action, including a TSL-1 predefined action, the value of the property for that task is updated to this most recent action. Note (see below) that basic events signifying predefined actions have equivalent forms as **performs** events that match the activator in Last_Done.

Property declarations and bodies are passed to the TSL-1 monitor when their declarative region in the Ada program is elaborated.

A property declared in the definition of a task type defines an attribute of each object of that type.

Example: Has_paid property of customer tasks.

```
task type CUSTOMER is
. . .
property HAS_PAID : BOOLEAN := FALSE
is
     when self performs PRE_PAY
          then set HAS_PAID := TRUE;
     when self performs PUMP_GAS
          then set HAS_PAID := FALSE;
end;
. . .
end CUSTOMER;
```

Commentary

Every customer object has a **HAS_PAID** attribute, tells whether the customer has pre-paid or not. The above definition is equivalent to defining a property **outside** the task type declaration with an extra parameter of type **CUSTOMER** (see the example in Section 5.1).

The value of a property may be referenced in a TSL-1 expression by a notation similar to a standard Ada function call. The property name is followed by a list of values of types corresponding to the list of TSL-1 types in the property declaration. Such an expression always denotes the *current* value of the property. For example, we can declare a global count of all active tasks. The definition uses the current value of the count property, as well as the predefined events, **activates** and **terminates**:

Example: The number of active tasks.

```
property ACTIVE_COUNT : INTEGER := 1
is
    when any activates
        then set ACTIVE_COUNT := ACTIVE_COUNT + 1;
    when any terminates
        then set ACTIVE_COUNT := ACTIVE_COUNT - 1;
end ACTIVE_COUNT;
```

The **ACTIVE_COUNT** expression on the right side of the assignments denotes the current value of the property.

TSL-1 provides predefined properties describing the status of Ada tasks, such as, **Running**, **Calling**, **Accepting**, and so on. Nearly all of the predefined properties can be defined by property declarations and bodies using the predefined basic events. (This is not possible in some cases because the set of TSL-1 predefined events is incomplete.)

Example: Definition of a predefined property.

```
property Calling(TASK) : BOOLEAN := FALSE is
    when ?T calls any
        then set Calling(?T) := TRUE;
    when any releases ?T
        then set Calling(?T) := FALSE;
end Calling;
```

If more than one **set** assignment to the same property is executed at the same event in the event stream, the TSL-1 computation is erroneous. The resulting value of the property will be one of the assigned values, but which one is not specified by TSL-1. If a type mismatch occurs during the execution of a **set** assignment, the TSL-1 computation is erroneous and the assignment is not executed.

Example: A well-defined property and an erroneous property.

```
property Count : NATURAL := 0
is
    when A then
        B and C set Count := Count + 1;
end Count;
```

Consider the stream interval: [A, B, C, B, C]

The values of Count are: 0, 0, 1, 1, 1

```
property Nonsense : TASK := null
is
    when ?T calls any or any calls ?T
        then set Nonsense := ?T;
end Nonsense;
```

Consider the stream interval: [T1 **calls** T2]

Commentary

Count will be activated by an A. Each activation of its update will result in incrementing its value at the first subsequent event that completes a match of B **and** C. Nonsense, on the other hand, will be activated twice by the calls event with different matches for ?T. These activations lead to two simultaneous **set** assignments with the different values, which is erroneous.

3.8 Guards

Often the occurrence of a stream event is important only in certain contexts. TSL-1 provides *guards* to specify context. A guard is a boolean condition built up from expressions (especially properties) using relational operators and boolean connectives. A guard is bound to a basic event by the reserved word **where** to form a *guarded event*. A Guarded event can match a stream event only when the guard condition is true.

Examples: Guarded events.

> OPERATOR **calls** CUSTOMERS(1) **at** CHANGE
> **where not** TERMINATED(CUSTOMERS(1))

> ?C **calls** ?P **at** START_PUMPING **where not** In_Use(?P)

> ?X **activates any where** ACTIVE_COUNT < 1000

Basic events are the only kind of events that can be guarded. Compound events and specifications are built up from guarded basic events, but may not be guarded themselves.

A guard denotes the value of a boolean expression at each stream event. Since a guard may change value as a new stream events occur, it is important to define an order in which matching and execution of updates are performed.

Whenever a new stream event is considered in attempting to match a guarded event, an order of matching of specifications and executing **set** assignments is defined as follows:

1. possible matching instantiations of guarded events are computed using the current values of properties to evaluate guards. This includes guarded events in specifications and in updates.
2. **set** assignments that are activated as a result of step 1 are executed as follows. All expressions on righthand sides of **set** assignments are first evaluated using the current values of properties. The assignments are then executed in some order that is not specified.

In the last example above, if the next stream event is an activation, and the current value of ACTIVE_COUNT is 999, the guarded event will be matched first — successfully. Then the value of ACTIVE_COUNT will be set to 1000.

During evaluation of a guard, Ada subexpressions are evaluated according to Ada rules. In particular, a placeholder in an expression must be replaced by a value associated with it in a (partially complete) matching before the expression is evaluated. If a placeholder is unbound by the matching, evaluation of the expression results in an erroneous TSL-1 computation. There is one exception to this rule. If an unbound placeholder occurs as an operand of an equality operator during guard evaluation, say, ?P = E, then the value of E is associated with the placeholder ?P and the guard is true. If both operands of an equality expression are unbound placeholders, the result is also erroneous.

3.9 Parameter Bindings

The optional parameter binding in basic events (the *call*, *accept*, *start rendezvous*, *end_rendezvous*, and *performs* categories) is given in the Ada named parameter form. The formal parameters of the Ada entry declarations or TSL-1 action declarations are used as parameter names. Only those parameters of interest need be named in the binding part of a basic event.

For example,
```
OPERATOR accepts ?C1 at PRE_PAY(AMOUNT => ?X1)
```
will match stream events in which the operator accepts prepayment from a task (in actuality, any customer), and ?X1 will be associated with the actual amount in each such event (AMOUNT is a formal parameter of PRE_PAY).

This extra context is sometimes important in specifying events accurately and reducing unwanted matching.

Examples of parameter bindings

```
?X calls OPERATOR at PRE_PAY(AMOUNT => ?Y) =>
?Z calls OPERATOR at PRE_PAY(AMOUNT => ?Y);

when OPERATOR accepts ?C1 at PRE_PAY(AMOUNT => ?X1)
    then not OPERATOR accepts ?C2 at PRE_PAY(AMOUNT => ?X2)
                                    where (?X2 < ?X1) =>
            ?P1 accepts ?C2 at START_PUMPING =>
            ?P1 accepts ?C1 at START_PUMPING
    until ?P2 accepts ?C1 at START_PUMPING;
```

Commentary

The first example matches pairs of events where customers prepay the same amount. The second example changes the race condition specification (Section 3.6) by adding an entry binding. This specification will be violated only if a customer who has paid less beats an earlier customer to the pump.

3.10 Optional Parts of Events and Specifications

Most classes of basic event have three or four components, some of which are optional. Optional parts are used to provide context. For example, a START RENDEZVOUS event has three optional parts, the first naming the task being accepted, the second naming the entry being accepted, and the third matching actual parameters of the entry call that is accepted.

Optional parts of basic events are given in the syntax summary. If they are omitted, the default is **any**. Thus

      OPERATOR **accepts** ?T   *means*   OPERATOR **accepts** ?T **at any**

and it will match any stream event in which the OPERATOR starts a rendezvous with a task — no matter which of its entries it has accepted.

The activation and termination events are optional parts of specifications. If an activator event is omitted, the specification becomes active when it is elaborated. If the terminator event is omitted, it will be terminated (and either satisfied or violated) when the elaborating thread of control exits that scope. These defaults for activation and termination of specifications allow us to express general constraints that apply throughout a scope. For example, one way to specify that a pump services only one customer at a time is to use the In_Use property in a specification as follows.

Example: Specification with default activation and termination.

    not ?X **accepts** ?Y **at** START_PUMPING **where** IN_USE(?X);

Commentary

There are no activation or termination events for this specification. By default it becomes active when it is elaborated, and it remains active until it is violated or control leaves the scope. Therefore this specification is equivalent to a constraint on all events in the stream while the scope is active — namely, for all stream events matching the event in the specification, the guard must be false. Notice that this example is not equivalent to the specification,

    ?X **accepts** ?Y **at** START_PUMPING **where not** In_Use(?X);

which will be satisfied the first time a stream event matches it and the (negated) guard is true, and will thereafter be inactive until re-elaboration.

There is also an option whereby specifications can be named using the label notation of Ada. Names of specifications cannot be referred to in TSL-1. They are intended for use by the TSL-1 implementation, e.g., for error messages.

3.11 Macro Definitions

Macro definitions allow commonly used parts of specifications to be named. This is intended to reduce the amount of repetition in constructing specifications, and to improve readability. A macro definition has the syntactic form,

    macro simple_name parameter-list **is** *compound*_event

where parameters are placeholders.

A *macro call* is placed in the position of an event in a specification. It consists of the macro name followed by a list of actual parameters, which may include placeholders. A macro call appears syntactically similar to a procedure call.

A macro call matches an interval of the event stream if there is an instantiation of its body that matches the interval. When a call is part of a larger event, some of its placeholders may be bound to values when the call is matched (as part of a partial instantiation of the larger event). That partial instantiation is applied to the macro body before it is matched. An unbound placeholder of the call will receive a value if and when the body matches an interval.

For example, two tasks are said to be *synchronized* at the start or end of any rendezvous between them. This is expressed by the macro:

```
macro SYNCHRONIZED(?S, ?T : TASK) is
    ?S accepts ?T or ?T accepts ?S
    or ?S releases ?T or ?T releases ?S
    or ?S activates ?T or ?T activates ?S;
```

This macro can be used to build more complex specifications. Assume, for example, that a variable is shared between tasks and that a TSL-1 action,

```
    action SHARED_VARIABLE_ACCESS;
```

is declared, and perform statements are inserted in the Ada program so that SHARED_VARIABLE_ACCESS is performed each time an Ada task accesses this variable. This implies that an event of the form,

```
    T performs SHARED_VARIABLE_ACCESS
```

will appear in the stream each time the variable is accessed by a task, T. Consider the specification,

```
    when ?A performs SHARED_VARIABLE_ACCESS
        then not ?B performs SHARED_VARIABLE_ACCESS
                where  ?B /= ?A
        before SYNCHRONIZED(?A, ?C) =>
            ?C performs SHARED_VARIABLE_ACCESS
```

The macro call, SYNCHRONIZED(?A, ?C), will match whenever its body matches after the formal parameter ?S is replaced by ?A and ?T is replaced by ?C in the body. When the activator matches, ?A receives a value. Then in order for the terminator sequence event to match, the macro call must match after ?A is replaced by its value. Each match of the macro body will a yield value for ?C, which is then substituted at the other position of ?C in the specification.

Suppose a stream event occurs in which a task U performs SHARED VARIABLE_ACCESS. A copy of the specification is activated. This activated copy will then continue to match events in which tasks start or end rendezvous with U thus yielding matches of the macro body, (and hence, of the call) until another task V performs SHARED_VARIABLE_ACCESS. At this point, either SYNCHRONIZED(U, V) has matched, and the specification is satisfied, or else it will be violated. The same event that terminates the currently active copy of this specification, will also activate another copy. Thus, this example specifies that if a task accesses the shared variable then it will synchronize with the next task to access that variable.

Macros may contain calls to themselves, in which case they are *recursive*. Recursive macros allow us to express arbitrary repetition of an event. This may be illustrated by constructing a macro to specify a sequence of synchronizing operations connection two tasks.

```
macro SYNC_PATH(?S, ?E : TASK) is
    SYNCHRONIZED(?S, ?E) or
    (SYNCHRONIZED(?S, ?T) => SYNC_PATH(?T, ?E));
```

A call to a recursive macro is an event in a specification. When the call is matched, this requires the macro body to match, which may result in an inner call being matched. For example, a call, SYNC_PATH(A, B,), will match any interval containing a sequence of events, each synchronizing the next pair of tasks in a chain, A, T1, . . . , Tn, B.

The semantics of recursive macro call requires some syntactic restrictions on where calls may be placed in the body of the macro. Essentially, a call must not be the first event in its own body. This prevents the matching process from getting into an infinite loop.

3.12 TSL-1 Task Specifications

Specifications of Ada tasks and task types may be constructed by placing TSL-1 declarations in the Ada task (type) declarations. The resulting specifications define those properties and actions of tasks that can be observed by other tasks, and specify relationships between those properties and actions. These augmented Ada/TSL-1 specifications can be viewed as abstract Ada task (type) definitions that specify the task bodies.

The meaning of TSL-1 task specifications depends on the meaning of TSL-1 constructs described in previous sections, and on some additional details which are described here. Our discussion for task types applies also to single task declarations.

Actions, properties, and TSL-1 specifications declared in an Ada task type declaration have the same scope as an entry of the task type. They are visible in TSL-1 declarations and statements in that scope, both inside and outside of the task type. They are said to *belong* to the task type, meaning that each task of the type is considered to possess those actions and properties, and conform to those specifications.

An action belonging to a task type may be performed (by perform statements) only inside the task body. Actions are used to declare the behavior of tasks that can be observed and responded to by other tasks.

A property belonging to a task type may refer in its definition to actions and properties that are visible, including those declared outside of the task type. Properties may be used to define abstract states of tasks of a task type.

A TSL-1 specification belonging to a task type is a visible promise. That is, one may assume that the behavior of each task of the type will obey the specification. It is also a constraint on the body of the task type. The body must be such that the behavior of any task of that type always satisfies the TSL-1 specification.

In the declaration of an Ada task type, the type name may be used in TSL-1 declarations. For example, it can be used to declare the type of parameters of actions and properties. The TSL-1 reserved word, **self** can be used to denote the executing thread of control, analogously to the Ada use of the type name in Ada task bodies.

Example: A TSL-1 specification of customers.

```
        task type CUSTOMER is

--+         action PRE_PAY(AMOUNT : in POSITIVE; ID : in PUMP);
--+         action PUMP_GAS( ID : in PUMP);
--+         action GET_CHANGE(AMOUNT : in NATURAL);

--+         property HAS_PAID(PUMP) : BOOLEAN := FALSE
--+         is
--+             when self performs PRE_PAY(ID => ?P)
--+                 then set HAS_PAID(?P) := TRUE;
--+             when self performs PUMP_GAS(ID => ?P)
--+                 then set HAS_PAID(?P) := FALSE;
--+         end;
```

```
--+        not (self performs PUMP_GAS(ID => ?P)
--+                  where not HAS_PAID(?P));

--+        when self performs PRE_PAY
--+             then self performs PUMP_GAS =>
--+                  self performs GET_CHANGE
--+             before self performs PRE_PAY;
```

end CUSTOMER;

Commentary

Customer tasks perform three visible actions. The property, HAS_PAID defines a state that is entered whenever a customer prepays, and is left whenever it pumps gas. The first specification is a promise that a customer is always in this state when it pumps gas. The second specification promises that the visible actions are performed in a certain order.

One implementation of this abstract Ada/TSL-1 task type specification is the Ada Customer task type in the Gas Station example, Section 5.1. The action, PRE_PAY is implemented as an entry call to the PRE_PAY entry of the OPERATOR. The action, PUMP_GAS is implemented as a sequence of two calls to the entries, START_PUMPING and FINISH_PUMPING, of a pump. The action, GET_CHANGE is implemented as a customer accepting a call to its CHANGE entry. The abstract customer specifications are transformed into TSL-1 constraints on the Ada program by replacing actions by their implementations (see specifications (1) in example 5.1).

4. Connected Events

In this section we describe briefly the concept of *connected events* in distributed systems in general, and then in Ada systems. The basic assumption about global event streams that represent computations of distributed Ada programs is stated using connected events. Consistency between TSL-1 specifications and Ada programs is then defined. Finally we introduce the concept of *robust specifications* based on connected events.

In an ideal model of a distributed system each thread of control executes its computation independently of any other thread of control. An *event* signifies (or results from) performance of an action in the distributed computation. Assume that an event has a unique name. Some pairs of events, A and B, may have to occur in any computation in a definite order, say A *occurs before* B, because one depends on the other. In this case we say that A and B are *connected* events, and we write $A < B$. Other pairs of events may occur in such a way that they have no significant order — they are *unconnected*.

Events resulting from the actions executed by a single thread of control are strictly ordered in a linear sequence. We will refer to this sequence as a *local event stream*. All pairs of events in a local event stream are connected.

Pairs of events, A, B occuring in separate local event streams may also be connected. In this case a popular model postulates connections as a result of the occurrence of a pair of *synchronizing events*, one in each stream. If A and B are a pair of events synchronizing two local streams, we write, $A \cong B$. The important property of synchronizing pairs of events is that they impose connections between events coming before and after them in their respective local streams. For example, in the streams,

$$stream1 : \ldots, e_1, A, e_2, \ldots$$
$$stream2 : \ldots, f_1, B, f_2, \ldots$$

if $A \cong B$ then $e_1 < B < f_2$ and $f_1 < A < e_2$.

In fact, we can use this property as the definition of a synchronizing pair of events. Namely, if A occurs in *stream1* and B occurs in *stream2* and satisfy,

for all e_1, e_2 in stream1: $e_1 < A < e_2 \rightarrow e_1 < B < e_2$,
for all f_1, f_2 in stream2: $f_1 < B < f_2 \rightarrow f_1 < A < f_2$,

then $A \cong B$.

Synchronizing pairs imply connectedness between other events according to the following axioms:

Axioms of connectivity.

> *(1)* **if $X < Y$ and $Y < Z$ then $X < Z$.**
> *(2)* **not ($X < Y$ and $Y < X$)**
> *(3)* **if $X \cong Y$ then $Y \cong X$.**
> *(4)* **if $X < Y$ and $Y \cong Z$ then $X < Z$.**
> *(5)* **if $X \cong Y$ and $Y < Z$ then $X < Z$.**

Consequently, the totality of events in a distributed computation is partially ordered by the relationship of *connectedness*.

A global event stream representing a computation of a system is obtained by merging each of the local event streams. It is required to be *consistent* with the connectedness relation. That is, if $X < Y$ then X is placed *before* Y in the global stream, and if $X \cong Y$ then both X and Y occur in the global stream in some order without any event of the corresponding two local streams between them. This consistency property will be satisfied if local event streams are merged into a global event stream so that,

1. the order of events in a local stream is preserved.
2. whenever a member A of a synchronizing pair of events occurs, neither A nor any later event from the local stream containing A is allowed to enter until the other member B of the pair has occurred; both A and B are then entered into the global stream, and merging of the two streams continues.

4.1 Connectedness in Ada Systems.

In distributed Ada systems the local event streams correspond to the individual tasks. Certain TSL-1 basic events correspond to Ada tasking actions. For example "S **calls** T **at** E" corresponds to S performing a call to an entry E of task T (i.e., an unconditional entry call). This means that some of the pairs of TSL-1 stream events are connected as a consequence of the semantics of Ada tasking. A calls event of the form, "T_1 **calls** T_2 **at** E" in the stream T_1 is connected to a start rendezvous event of the form, "T_2 **accepts** T_1 **at** E" in the stream T_2. Generally, there are many similar calls and start rendezvous events in the two streams. Only some of these pairs are required to be *related* as a consequence of Ada semantics, and thereby become connected. The connected pairs of calls and start rendezvous events are defined by induction as follows:

(A) Connection of Calls and Accepts

Consider events in the order of the local streams T_1 and T_2. The first calls event, "T_1 **calls** T_2 **at** E" in the stream T_1 that is not already related to an event, is related to the first start rendezvous event, "T_2 **accepts** T_1 **at** E" in the stream T_2 that is not already related to an event, and conversely. If e_1 is a calls event and e_2 is the related start rendezvous event, then they are connected in the order, $e_1 < e_2$. Furthermore, for any other event f in the same local stream as e_1, if $e_1 < f$ then $e_2 < f$.

Other pairs of classes of TSL-1 basic events satisfy connection axioms similar to (A) — e.g., calls and releases events.

Pairs of events occuring in an Ada computation may be connected because they result from actions performed in a single local stream, or because they are connected by the semantics of Ada as expressed by assumptions such as (A). Such pairs of events are said to be *directly connected*. Other pairs of events may be connected as a consequence of the general axioms of connectivity applied to the directly connected pairs that occur in a computation. In the latter case, the events are said to be *indirectly connected*.

A pair of events, P, is indirectly connected as a result of the way the underlying Ada program is constructed — i.e., the program executes a directly connected pair between the members of P. Indirect connections may be specified in TSL-1 as shown in Example 5.2.

Examples: Connected events in Ada systems.

(1) $stream1$: T_1 **calls** T_2 at E -- *entry call event,*
     $stream2$: T_2 **accepts** T_1 at E -- *the related start rendezvous,*

are directly connected, if they are related as a consequence of Ada semantics. Related pairs of events of these kinds must always happen in the same order in every computation.

(2) $stream1$: T_1 **calls** T_2 at E -- *an event in local stream1*
     $stream1$: T_1 **calls** T_3 at E -- *an event in local stream1*

are directly connected because they occur in the same local stream.

(3) $stream1$: T_1 **calls** T_2 at E -- *an event in local stream1*
     $stream3$: T_3 **calls** T_2 at E -- *an event in local stream3*

are not directly connected. They may be indirectly connected if a pair of connected events occurs in *stream1* and *stream3* in such a way that the two **call** events are separated by the pair. Generally, they are not connected.

Basic Assumption of TSL-1.

The basic assumption of TSL-1 is that the order of events in a global event stream representing a computation of the underlying Ada program is *consistent* with the partial order of connected predefined TSL-1 basic events that is generated in that computation.

It should be noted that the set of predefined TSL-1 events does not contain synchronizing pairs of events. Instead those pairs of events in different local streams that are directly connected are defined inductively by axioms such as (A) above. In order to satisfy the basic assumption, a global event stream must be constructed by merging local streams so as to satisfy axioms (A) and the general connectivity axioms. Also, the execution of Ada constructs such as conditional and timed entry calls is not represented in the TSL-1 predefined events. It is assumed that such calls are not executed in the underlying Ada system. The definition of connections between these kinds of events and corresponding rendezvous events can be achieved only by adding new predefined events.

4.2 Consistency of TSL-1/Ada programs

A TSL-1 specification is *satisfied by an event stream* if every one of its activations is satisfied.

An Ada program is *consistent* with a TSL-1 specification if for every computation of the program, the TSL-1 specification is satisfied by all global event streams that represent the computation according to the basic TSL-1 assumption.

The concept of consistency between a program and a specification, as we have defined it here, implies that the runtime behavior of the program never violates the specification. We will also say that the underlying Ada program is *correct* if it is consistent with all of its TSL-1 specifications.

Runtime monitoring, which we have used to define the semantics of TSL-1, provides a technique for detecting violations, but cannot be used to establish consistency. Consistency may be proved by mathematical proof based on proof rules for TSL-1. Discussion of alternative definitions of the semantics of TSL-1 by proof rules is beyond the scope of this paper.

4.3 Robust TSL-1 Specifications

A TSL-1 specification of a program is *robust* if and only if for any computation it has the same result (satisfied or violated) on all global event streams representing that computation.

Any two global streams representing the same computation of the underlying program can only differ in the order of events that are not connected. Therefore a specification is robust if every pair of its basic events for which an order is specified can only match connected stream events.

A specification that is consistent with a program is robust, but not conversely. The job of determining consistency is made easier if it is known that a specification is robust, for then inconsistencies cannot be hidden by some event streams and exposed by others. Therefore it is important to use specifications

A subclass of robust specifications are those such that every pair of basic events in the specification for which an order is specified can only match directly connected stream events. These are called *directly robust* specifications. This kind of robustness is independent of the underlying Ada program. It is often easy to check if a specification is robust in this sense.

Let us consider two examples.

Example of a TSL-1 sequence that is not directly robust.

```
CUSTOMERS(1) calls OPERATOR at PRE_PAY     =>
CUSTOMERS(2) calls OPERATOR at PRE_PAY     =>
CUSTOMERS(2) calls PUMPS at START_PUMPING =>
CUSTOMERS(1) calls PUMPS at START_PUMPING;
```

Commentary

Consider events in the local streams generated by the customer tasks:

CUSTOMERS(1)

e_1 : CUSTOMERS(1) **calls** OPERATOR **at** PRE_PAY

e_2 : CUSTOMERS(1) **calls** PUMPS **at** START_PUMPING

CUSTOMERS(2)

e_3 : CUSTOMERS(2) **calls** OPERATOR **at** PRE_PAY

e_4 : CUSTOMERS(2) **calls** PUMPS **at** START_PUMPING

As a result of direct connectivity in local streams, $e_1 < e_2$ and $e_3 < e_4$. If we assume that the underlying program does not generate any other connected pairs, then only these two pairs are connected. The four events may appear in the global stream in different orders, including

e_1, e_2, e_3, e_4 and e_1, e_3, e_4, e_2 and e_3, e_4, e_1, e_2.

The TSL-1 sequence will match the second stream but not the others.

Therefore it is possible to construct two global event streams that represent the same computation of the program such that the TSL-1 sequence will match one stream and not the other. Consequently the fact that it matches one stream does not indicate that it is satisfied by that computation of the underlying program.

Example of a robust TSL-1 specification.

```
OPERATOR accepts CUSTOMERS(1) at PRE_PAY        =>
OPERATOR accepts CUSTOMERS(2) at PRE_PAY        =>
PUMPS(1) accepts CUSTOMERS(2) at START_PUMPING =>
PUMPS(1) accepts CUSTOMERS(1) at START_PUMPING;
```

Commentary

Ada semantics as expressed by (A) imply that this sequence is a *robust* TSL-1 sequence as a consequence of direct connections. Its robustness is independent of the underlying program. It is not, however, a directly robust specification.

Any sequence of four events in the global stream, e_1, e_2, e_3, e_4, that matches the TSL-1 sequence must be connected in the order, $e_1 < e_2 < e_3 < e_4$. To see this, consider the four local streams generated by the operator, two customers, and pumps (1). Events e_1, e_2, e_3, e_4 must occur in their respective local streams in the order indicated as a consequence of the basic TSL-1 assumption. Events e_5, e_6, e_7, e_8 are chosen to be the calls events related to them under connection of Calls and Accepts (A):

```
OPERATOR e₁ : OPERATOR accepts CUSTOMERS(1) at PRE_PAY,
         e₂ : OPERATOR accepts CUSTOMERS(2) at PRE_PAY,

PUMPS(1)
     e₃ : PUMPS(1) accepts CUSTOMERS(2) at START_PUMPING,
     e₄ : PUMPS(1) accepts CUSTOMERS(1) at START_PUMPING,

CUSTOMERS(1)
     e₅ : CUSTOMERS(1) calls OPERATOR at PRE_PAY,
     e₆ : CUSTOMERS(1) calls PUMPS(1) at START_PUMPING,

CUSTOMERS(2)
     e₇ : CUSTOMERS(2) calls OPERATOR at PRE_PAY,
     e₈ : CUSTOMERS(2) calls PUMPS(1) at START_PUMPING,
```

Although e_5, e_6 and e_7, e_8 are connected pairs of events, we do not know their order of occurrence (e.g., if a customer calls the operator and uses the pump more than once, how do we know if e_5 and e_6 are events in the same transaction?).

(1) Assume that $e_7 < e_8$.
     Then: $e_7 < e_2$ and $e_2 < e_8$ (axiom (A) applied to e_7, e_2, and e_8), $e_8 < e_3$ (axiom (A) applied to e_8 and e_3).
     Consequently we have the connections,
     $e_1 < e_2, e_2 < e_8, e_8 < e_3, e_3 < e_4,$
     from which our claim follows by transitivity.

(2) Assume that $e_8 < e_7$.

Then: $e_8 < e_3$ and $e_3 < e_7$ (axiom (A) applied to e_8, e_3, and e_7), $e_7 < e_2$ (axiom (A) applied to e_7 and e_2).
(Consequently $e_3 < e_2$ by transitivity.)

This means that e_3 is connected to e_2 and must occur in the global stream in the order $e_3 < e_2$ (by the basic assumption). The hypothesis that e_2 occurs before e_3 in the global stream is contradicted. Therefore assumption (2) is false.

It is important to formulate guidelines for constructing TSL-1 specifications that are robust, and to be able to test them for robustness. Sufficient tests that imply robustness based on direct connections can be defined. In general however, these tests are too stong, since they do not take into account other stream events which may be indirectly connected. We define a sufficient test below.

(1) Basic events in a specification.

A pair of TSL-1 basic events *passes the robustness test* if the events occur in a compound event or specification and satisfy one of the following two properties: *(1)* every pair of stream events that they match is from the same local event stream, or *(2)* every pair of stream events that they match is connected by assumptions such as (A) imposed by Ada semantics.

Case (1) is satisfied if they have a common task or task placeholder that must match the executing thread of control for those events. For case (2), a sufficient test is that the pair of TSL-1 events must occur within a context (TSL-1 compound event or specification) so that the semantics of TSL-1 implies that they must match pairs of stream events that are *related* under a connection assumption.

Definition

A pair of sets of TSL-1 events *passes the robustness test* if every pair of events consisting of an event from the first set and an event from the second set, passes the robustness test.

Definition

Robustness of a TSL-1 compound event is tested by associating with each compound event two sets of TSL-1 events, the *head* and the *tail* for that compound event. The head or tail of a guarded event is simply the guarded event itself. The head (tail) of a sequence is the head (tail) of the first (last) component of the sequence. The head (tail) of a one_of or all_of compound event is the union of the heads (tails) of its component events. The head (tail) of an iterated event is the head (tail) of the component event to which the iterator applies.

(2) TSL-1 compound events

A sequence compound event *passes the robustness test* if every component passes the test, and each pair of sets consisting of the head of a component and the tail of the previous component passes the test. A one_of or all_of event passes the test if all of its components pass the test.

(3) TSL-1 specifications

A TSL-1 specification *passes the robustness test* if *(1)* all three component events, the activator, the body, and the terminator pass the test, and *(2)* the pairs of sets consisting of the tail of the activator and the head of the body, the tail of the activator and the head of the terminator, and the tails of the body and terminator, all pass the robustness test for pairs of sets.

It is easy to write a TSL-1 semantic analyser that checks for robustness under this test. It would issue a warning for any specification that failed the test, indicating that robustness must depend on other events that are not mentioned in the specification or else specifications for user defined actions.

5. Examples

5.1 Simulation of an Automated Gas Station

The gas station contains tasks representing the station operator, the customers, and the pumps. A customer arrives at the station, prepays the operator, pumps gas, receives change from the operator, and leaves. The operator accepts prepayments, activates pumps, computes charges and gives customers their change. A queue package is used to queue prepaid customers if there are overlapping requests for pumps. The pumps are used by the customers, and report the amount pumped to the operator.

The example illustrates the use of TSL-1 to specify task interactions in a completed Ada Program. The basic Ada declarations, some of the TSL-1 specifications, and the Ada bodies are shown. Only predefined TSL-1 events are used, so there are no user-defined actions. It can be easily seen, intuitively speaking, that all but one of the TSL specifications are satisfied by the bodies.

```
with Queue_Class, Random;
procedure GAS_STATION is

    task OPERATOR is
        entry PRE_PAY(AMOUNT      : POSITIVE;
                      PUMP_ID      : POSITIVE;
                      CUSTOMER_ID : POSITIVE);

        entry CHARGE(AMOUNT  : POSITIVE;
                     PUMP_ID : POSITIVE);
    end OPERATOR;

    task type CUSTOMER is
        entry GET_CUSTOMER_ID(NUMBER : POSITIVE);
        entry CHANGE(AMOUNT : NATURAL);
    end CUSTOMER;

    task type PUMP is
        entry GET_PUMP_ID(NUMBER : POSITIVE);
        entry ACTIVATE(LIMIT : POSITIVE);
        entry START_PUMPING;
        entry FINISH_PUMPING(AMOUNT_CHARGED : out NATURAL);
    end PUMP;

-- 1. specification of constraints on customers.
    --+ property HAS_PAID(CUSTOMER, PUMP) : BOOLEAN := FALSE
    --+ is
    --+     when OPERATOR releases ?C from
    --+                              PRE_PAY(PUMP_ID => ?P)
    --+         then set HAS_PAID(?C, ?P) := TRUE;
    --+     when ?C calls ?P at FINISH_PUMPING
    --+         then set HAS_PAID(?C, ?P) := FALSE;
    --+ end HAS_PAID;

    --+ not ?C calls ?P at START_PUMPING
    --+     where not HAS_PAID(?C, ?P);

-- a customer's protocol.
    --+ when ?C calls OPERATOR at PRE_PAY
    --+     then ?C calls ?P at START_PUMPING  =>
    --+          ?C calls ?P at FINISH_PUMPING =>
    --+             ?C accepting CHANGE
    --+         before ?C accepts OPERATOR at CHANGE;
```

```
--  2. specification of a pump's protocol.
    --+ when ?P accepts OPERATOR at ACTIVATE
    --+      then ?P accepts ?C at START_PUMPING  =>
    --+           ?P accepts ?C at FINISH_PUMPING =>
    --+           ?P calls OPERATOR at CHARGE
    --+      before ?P accepts OPERATOR at ACTIVATE;

--  3. constraint on the operator.
    --+ when OPERATOR accepts ?P at CHARGE
    --+      then OPERATOR calls ?C at CHANGE
    --+      before OPERATOR accepting;

--  4. specification guarding against races for a pump.
    --+ when OPERATOR accepts ?C1 at PRE_PAY
    --+      then not OPERATOR accepts ?C2 at PRE_PAY  =>
    --+               ?P1 accepts ?C2 at START_PUMPING =>
    --+               ?P1 accepts ?C1 at START_PUMPING
    --+      until ?P2 accepts ?C1 at START_PUMPING;

    Num_Pumps     : constant Integer := 3;
    Num_Customers : constant Integer := 5;

    type Element is
        record
            Prepay_Amount, Customer_Id : POSITIVE;
        end record;

    package Customers_At_Pumps is new Queue_Class(Element);
    use Customers_At_Pumps;
    Pump_Queues : array (1 .. Num_Pumps) of
                                        Customers_At_Pumps.Queue;
    Customers   : array (1 .. Num_Customers) of Customer;
    Pumps       : array (1 .. Num_Pumps)      of Pump;

    procedure Drive_In(MONEY : out POSITIVE;
                       PUMP  : out POSITIVE) is ...
        -- called by customers to randomly decide amount of money and
        -- choice of pump.

    function Get_Charges(Limit : NATURAL) return NATURAL is ...
        -- called by pumps to randomly decide how much a customer has
        -- pumped.
```

```
task body Operator is
    Customer_Info   : Element;
    Amount_Pumped   : NATURAL;
    Idle_Pump       : POSITIVE;
    Top             : Integer;
begin
    loop
        select
            accept Pre_Pay(Amount      : POSITIVE;
                           Pump_Id     : POSITIVE;
                           Customer_Id : POSITIVE) do
                if Is_Empty(Pump_Queues(Pump_Id)) then
                    Pumps(Pump_Id).Activate(Amount);
                end if;

                Enqueue(Item => (Amount, Customer_Id),
                        Onto => Pump_Queues(Pump_Id));
            end Pre_Pay;
        or
            accept Charge(Amount  : POSITIVE;
                          Pump_Id : POSITIVE) do
                Dequeue(Item    => Customer_Info,
                        Off_Of => Pump_Queues(Pump_Id));
                Amount_Pumped := Amount;
                Idle_Pump := Pump_Id;
            end Charge;
            if not Is_Empty(Pump_Queues(Idle_Pump)) then
                Top := Front_Of(Pump_Queues
                        (Idle_Pump)).Prepay_Amount;
                Pumps(Idle_Pump).Activate(Top);
            end if;
            Customers(Customer_Info.Customer_Id).Change
                (Customer_Info.Prepay_Amount -
                                        Amount_Pumped);
        or
            terminate;
        end select;
    end loop;
end Operator;

task body Customer is
    My_Number     : POSITIVE;
    My_Money      : NATURAL;
    Pump_Number   : POSITIVE;
    Amount_Pumped : NATURAL;
```

```
begin
    accept Get_Customer_Id(Number : POSITIVE) do
        My_Number := Number;
    end Get_Customer_Id;

    Drive_In(My_Money, Pump_Number);
    Operator.Pre_Pay(My_Money, Pump_Number, My_Number);
    Pumps(Pump_Number).Start_Pumping;
    Pumps(Pump_Number).Finish_Pumping(Amount_Pumped);
    accept Change(Amount : NATURAL);
end Customer;

task body Pump is
    My_Number        : POSITIVE;
    Current_Charges  : NATURAL;
    Amount_Limit     : POSITIVE;
begin
    accept Get_Pump_Id(Number : POSITIVE) do
        My_Number := Number;
    end Get_Pump_Id;
    loop
        select
            accept Activate(Limit : POSITIVE) do
                Amount_Limit := Limit;
            end Activate;
            accept Start_Pumping;
            accept
                Finish_Pumping
                    (Amount_Charged : out NATURAL) do
                Current_Charges :=
                                Get_Charges(Amount_Limit);
                Amount_Charged := Current_Charges;
                Operator.Charge
                    (Current_Charges, My_Number);
            end Finish_Pumping;
        or
            terminate;
        end select;
    end loop;
end Pump;
```

```
begin
    for I in 1 .. Num_Pumps loop
        Pumps(I).Get_Pump_Id(I);
    end loop;

    for I in 1 .. Num_Customers loop
        Customers(I).Get_Customer_Id(I);
    end loop;

end GAS_STATION;
```

Commentary:

The TSL-1 specifications are stated at a global level so that all threads of control are constrained by them.

The specifications in group (1) constrain customers. The first specification requires that a customer task always prepays for a pump before pumping. The specification allows a customer to pump more than once (although this implementation does not do that). Therefore, a property, HAS_PAID is defined, which is true every time the customer pays, and reverts to false the moment that customer finishes pumping gas. The negated specification, "a customer does not activate a pump when that task has not paid for it", is activated on scope entry, and remains active until scope exit. The customer protocol specification defines the order in which a customer will interact with the other tasks during a transaction.

Specification (2) is a similar protocol defining the order in which a pump carries out its interactions with other tasks during a single transaction.

Specification (3) requires the operator to give customers change promptly. It does not, however, specify the change is given to the "correct" customer. We have omitted a protocol for the operator.

Specification (4) is the constraint against races for a pump. It requires that whenever a customer prepays, that task will get to start pumping before another customer makes a later prepayment and uses the same pump as the first customer.

Specifications (1) - (3) are robust according to the sufficient tests given in 4.3. Robustness of specification (4) follows from the robustness of the example sequence in Section 4.3 by similar arguments to those used to prove robustness of that sequence. It does not satisfy the sufficient tests for robustness.

The TSL-1 specifications given here illustrate the kinds of specifications that could be used to monitor the Ada program for errors. They can be compared with the task bodies. It should be intuitively clear that each of the specifications (1) - (3) are always satisfied by any computation, and hence are consistent with the program. However, the race constraint (4) may be violated on some computations and is therefore inconsistent with the program.

There are three points of note. First, equivalent protocol specifications can be placed in the task declarations, as in the case of Pumps:

task type PUMP **is**

      entry GET_PUMP_ID(NUMBER : **in** POSITIVE);
      entry ACTIVATE(LIMIT : **in** POSITIVE);
      entry START_PUMPING;
      entry FINISH_PUMPING(AMOUNT_CHARGED : **out** POSITIVE);

--+ **when self accepts** OPERATOR **at** ACTIVATE
--+   **then self accepts** ?C **at** START_PUMPING =>
--+   **self accepts** ?C **at** FINISH_PUMPING =>
--+   **self calls** OPERATOR **at** CHARGE
--+   **before self accepts** OPERATOR **at** ACTIVATE;

end PUMP;

This specification will constrain each PUMP object exactly as the previous global one; **self** designates an executing thread of control of the task body. (See also Section 3.12.)

Secondly, because of the TSL-1 semantics of =>, the TSL-1 protocols do not express "immediacy". A pump may satisfy the specification while doing other actions between the specified ones. For example, it could issue spurious calls to the operator just so long as the operator does not activate it again until it has completed a proper sequence. The protocol can be strengthened so that a pump must execute exactly the specified sequence of actions and no others; we use a property NEXT that counts pump actions:

task type PUMP **is**
    . . .   *-- entries as before.*
--+ **property** NEXT : NATURAL := 0
--+ **is**
--+   **when** (**self accepts any or self calls any**)
--+   **then set** NEXT := NEXT + 1;
--+ **end** NEXT;

```
--+  when self accepts OPERATOR at ACTIVATE where ?I = NEXT
--+       then self accepts ?C at START_PUMPING
--+            where NEXT = ?I + 1 =>
--+            self accepts ?C at FINISH_PUMPING
--+            where NEXT = ?I + 2 =>
--+            self calls OPERATOR at CHARGE
--+            where NEXT = ?I + 3
--+       before self accepts OPERATOR at ACTIVATE;

end PUMP;
```

Thirdly, the TSL-1 specifications in this example can be regarded as expressing both *liveness* and *safety* requirements. Typically, liveness is a requirement that something must happen. Safety is a requirement that something must not happen.

The two protocols express *bounded liveness*. Once a protocol is activated the specified sequence must happen before the terminator happens or the Ada scope is exited. This means that if a termination event will happen, the TSL-1 protocols express that sequences of events must happen. In practice, if some part of the program is doing something (e.g., the clock is ticking), liveness can always be expressed as bounded liveness by choosing an appropriate terminator. The other TSL-1 specifications, e.g., the constraint on customers to prepay, and the constraint against races, express safety properties.

5.2 A Byzantine General's Problem

This example illustrates the use of action declarations and TSL-1 specifications to specify task communications prior to implementation. The actions hide the actual details of how the communication takes place. The Ada program is divided informally into two parts, a *top level design* and a *simulation*, to distinguish between the abstract actions and one of many possible implementations.

The basic declarations show an abstract program in which actions are not implemented by Ada text. Dependency of one action on another is specified in TSL-1. The main behavioral requirement of the program is also specified. At this level, a designer may wish to write a "quick simulation" (or "rapid prototype") to establish if the specifications can be satisfied.

The *simulation* contains the rest of the Ada program. It represents a very naive way of implementing the General, the Soldiers, and the actions; it may not be the final implementation of choice. Each action is interpreted in this implementation by TSL-1 **perform** statements that indicate when the action is executed.

For brevity, we have not given all of the TSL-1 specifications. Protocols for Soldiers, for example, are omitted. Only those specifications dealing with dependencies between actions of different tasks and with reliability are shown here.

-- *Top level actions and specifications.*

```
    generic
        Max_Soldiers : Positive;
    procedure Byzantine;

    procedure Byzantine is

        type Move is (Advance, Retreat);

        task General is
--+         action Order(X : Move; S : Task);
        end General;

        task type Soldier is
            entry  Receive_Identity(I : Integer);
--+         action Receive_Order(X : Move);
--+         action Send_Copy(A : Move; L : Soldier);
--+         action Act(X : Move);
        end Soldier;
```

-- *Specification of relationship between actions*

```
--+     property Ready(Soldier) : Boolean := False
--+     is
--+         when General performs General.Order(S => ?L)
--+             then set Ready(?L) := True;
--+         when ?L performs Soldier.Receive_Order
--+             then set Ready(?L) := False;
--+     end;

--+     not ?L performs Soldier.Receive_Order
--+                         where not Ready(?L);

--+     not General performs General.Order(S => ?L)
--+                             where Ready(?L);

--+     not ?L2 performs Soldier.Act   =>
--+         any performs Soldier.Send_Copy(L => ?L2);
```

```
--    Specification of Reliability and Agreement.

--+        property Reliable(Soldier) : BOOLEAN := TRUE is
--+            when ?L performs Soldier.Receive_Order(X => ?M)
--+                then ?L performs Soldier.Send_Copy(A => ?M1)
--+                    where ?M1 /= ?M
--+                        then set Reliable(?L) := FALSE;
--+        end Reliable;

--+ <<AGREEMENT>>
--+        not ?L1 performs Soldier.Act(A => ?M1)
--+            where Reliable(?L1) =>
--+            ?L2 performs Soldier.Act(A => ?M2)
--+            where Reliable(?L2) and ?M2 /= ?M1;

--    Simulation of top level specification.
            type Army is array (Positive range <>) of Soldier;
            type Mail is array (Positive range <>) of Move;
            Soldiers            : Army(1 .. Max_Soldiers);
            Mailboxes, Actions : Mail(1 .. Max_Soldiers);
            Communications      : array (1 .. Max_Soldiers)
                of Mail(1 .. Max_Soldiers);

            function Decide(X : Move) return Move is ... end;
            function Summarize(X : Mail) return Move is ... end;

--    Byzantine bodies.
            task body General is
                M : Move := Decide(Advance);
            begin
                for I in 1 .. Max_Soldiers loop
--+                 perform Order(M, Soldiers(I));
                        Mailboxes(I) := M;
                end loop;
            end General;

            task body Soldier is
                My_Id : Integer;
                A, Copy : Move;
            begin
                accept Receive_Identity(I : Integer) do
                    My_Id := I;
                end Receive_Identity;
                delay 5.0;
                A := Mailboxes(My_Id);
```

```
--+          perform Receive_Order(A);
             for J in 1 .. Max_Soldiers loop
                 if J /= My_Id then
                     Copy := Decide(A);
--+                  perform Send_Copy(Copy, Soldiers(J));
                     Communications(J)(My_Id) := Copy;
                 end if;
             end loop;
             delay 5.0;
             Actions(My_Id) :=
                 Decide(Summarize(Communications(My_Id)));
--+          perform Act(Actions(My_Id));
         end Soldier;

     begin
         for J in 1 .. Max_Soldiers loop
             Soldiers(J).Receive_Identity(J);
         end loop;

     end Byzantine;
```

Commentary

The actions are declared in the task specifications. The General, for example, is specified to perform Order actions. Soldiers are specified to perform three actions, receiving an order, sending copies of orders, and acting. The TSL-1 specifications constrain performance of these actions.

The property Ready and the two specifications that follow it together *relate* pairs of Order and Receive_Order events resulting from actions performed by the General and the Soldiers respectively. Consider any global stream that satisfies these specifications. Two stream events are related under this property if they match the two activators in a single instantiation of the property body, and if in addition, they are the earliest events in their respective local streams that are not already related to other events under the property. The first specification constrains the event stream so that an Order event must occur *before* a related Receive_Order event. The second specification constrains the event stream so that once a single Order event has occurred, then the related Receive_Order event must occur before another Order event. Thus the specifications require pairwise occurrence of related events. The *related pairs* are defined inductively over the event stream. In any implementation that is consistent with these constraints, related pairs of Order and Receive_Order events will always be observed in this order in every event stream. Related pairs of these events are therefore *connected* if the implementation is consistent with the specifications.

The third specification requires that a soldier does not act before a comrade sends a copy of an order to him. In an implementation that is consistent with this specification, all events of a soldier sending a copy to another soldier must precede an event resulting from the receiving soldier performing the action, Act. Consequently, Copying and Acting must be connected otherwise some event stream would contain such events in the wrong order.

Notice that the first and second specifications constrain corresponding pairs of Order and Receive_Order actions in the case when the General send more than one order to a Soldier, and a Soldier performs Receive Order many times. The third specification simply constrains all Send Copy actions to a Soldier to be performed before it Acts.

A protocol requiring soldiers to send copies of their orders to comrades is not given here, but it would be expressed similarly to protocols in the Gas Station.

Reliability is defined as a property possessed by Soldiers. It is set to true initially — all soldiers are assumed reliable until their actions prove otherwise. *Agreement* between reliable soldiers is expressed by a very simple specification that will be violated if there is a single event of a reliable Soldier performing an act *and* a single event in which a reliable Soldier performs a different act. Actually, the specification is expressed as a sequence instead of a conjunction. This is because the current version of TSL-1 does not allow a boolean expression as a specification.

These specifications are all placed in the top level declarative region. They will be activated when the main program elaborates, and remain active until violated or else the main program terminates.

The remainder of the Ada program is informally called a *simulation*. It implements the actors (General and Soldiers) and the actions. TSL-1 perform statements are used to express a correspondence between the actions and the simulation. A perform statement is placed so as to assert when an action is being performed.

Delay statements are included as a crude way to ensure that most computations will lead to event streams that satisfy the specifications. This simulation is not consistent with the specifications. In fact, it is actually an erroneous Ada program according to the Ada manual. The events that are specified to be connected are not connected. If, for example, the processors are slow enough, the 5 second delays will not ensure that actions are performed in the specified order. A consistent simulation must perform other actions resulting in events that connect the performance of the actions of the General and the Soldiers, and the Soldiers with each other.

The Byzantine procedure may be instantiated in a simple test program. If this program is processed by the TSL-1 compiler and then compiled (by an Ada compiler) and executed, it will generate a stream of events. This stream will consist mainly of performs events for user-defined actions. It will be monitored for consistency with the TSL-1 specifications.

A possible initial interval of an event stream for this program is:

```
main activates General,
main activates Soldiers(2),
main activates Soldiers(1),
        ...           -- activation of all other tasks.
main calls Soldiers(1) at Receive_Identity(I => 1),
General performs Order(Advance, Soldiers(1)),
Soldiers(1) accepts main at Receive_Identity(I => 1),
Soldiers(1) releases main from Receive_Identity(I => 1),
main calls Soldiers(2) at Receive_Identity(I => 2),
General performs Order(Advance, Soldiers(2)),
Soldiers(2) accepts main at Receive_Identity(I => 2),
Soldiers(2) releases main from Receive_Identity(I => 2),
Soldiers(2) performs Receive_Order(Advance),
Soldiers(2) performs Send_Copy(Advance, Soldiers(1)),
Soldiers(2) performs Send_Copy(Advance, Soldiers(3)),
Soldiers(1) performs Receive_Order(Advance),
Soldiers(1) performs Send_Copy(Advance, Soldiers(2)),
main calls Soldiers(3) at Receive_Identity(I => 3),
        ...
```

It is possible to use perform statements to express the correspondence between actions and implementation because the simulation in this example is very simple. In general, actions may involve many observable events so that their execution leads to overlapping sequences of events. In TSL-2 a more powerful action body construct will be provided to deal with such cases.

This example also illustrates the need for proof rules for TSL-1. Tasking programs can sometimes run for a very long time without revealing inconsistencies. TSL-1 provides a means of specifying behavior. The current runtime system however will only report inconsistencies when they occur, together with the stream that caused a violation of a specification. A proof system would hopefully allow us to prove: (1) that the specifications imply a connection between certain pairs of Order and Receive_Order events, and (2) that the performance of these actions do not result in connected events in this simulation.

When a simulation satisfies the TSL-1 specifications, the designer may choose to continue developing the program in several ways: (1) a more sophisticated implementation may be needed for the actions so there is no shared data between tasks, (2) further specifications may be imposed, (3) a module such as the General may be expanded in detail that does not affect communication with the army, such as a "joint chiefs of staff" that debate internally. TSL-2 will contain constructs to support these kinds of refinements within a structured discipline so that consistency between design and simulation, when established at one level, is not lost in further development.

6. Relationship to Other Work

Some of the concepts in TSL-1 have appeared in one form or another in earlier papers on specifying both the sequential and concurrent aspects of programs. We mention briefly some important related works. First, sequences and recursive macros in TSL-1 specifications are closely related to path expressions for defining execution sequences to synchronize processes [3]. Secondly, the idea of using sequences to specify concurrent processing has often been proposed in the research literature, Early pioneering examples being Gilles Kahn [10], and Ole-Johan Dahl [13]. Similar ideas have also been applied more recently in the event description language, EDL described by Bates and Wileden in [2]. The query aspects of TSL-1, as embodied in the concept of matching, has something of the flavor of Carl Hewitt's Planner [9], and later generalizations of Planner such as Prolog [4]. Application of Prolog to checking for deadlock errors in the history of Ada tasking programs has been investigated by Carol LeDoux in [12]. Finally, an operational semantics for TSL-1 is given in a separate paper by mapping TSL-1 constructs into event/token graphs, and defining the matching process in terms of operations on those graphs. Event/token graphs are closely related to Petri nets [14]. We did not investigate the use of Petri nets directly in defining TSL-1 semantics.

A. TSL Augmented Syntax

This appendix presents the syntax used in the prototype implementation of TSL. The prototype is a subset implementation in that it does not allow all operations defined for discrete types to be used in TSL declarations and specifications.

This appendix is intended not only to present the TSL syntax, but also to convey many of TSL's semantic restrictions. The semantic restrictions are primarily represented by an italic prefix added to syntactic elements. The italic prefixes used are *task*, *entry*, *discrete*, *integer*, *boolean*, *task_type*, *action*, *property*, and *macro*.

The prefixes *task*, *entry*, *discrete*, *integer*, *boolean*, and *action* indicate that the item must be of the appropriate type or type class. The prefixes *task*, *entry*, *task_type*, and *action* denote the corresponding predefined TSL type; *discrete* denotes the Ada type class; *integer* and *boolean* denote the predefined Ada types.

The precedence of operators used in TSL is the same as the precedence of the same operator in Ada. The only operators used in TSL which are not also available in Ada are those defined for compound events. The precedence of these operators, in increasing precedence, is:

or   and   =>

The nonterminal "value" stands for a value of some type. It is used in the syntax where the type of a value can not be determined syntactically.

Values, Names, and Expressions

```
value           ::= task_name | entry_name | task_type_name
                    | action_name | discrete_exp
task_name       ::= task_object | task_placeholder
                    | task_property | main | self | null
entry_name      ::= entry_identifier [(discrete_exp)]
                    | entry_placeholder | entry_property | null
action_name     ::= action_name | action_placeholder
                    | action_property | calls | accepting
                    | accepts | releases | activates | performs
                    | terminates | null
task_type_name  ::= task_type | task_type_placeholder
                    | task_type_property | null
placeholder     ::= ?simple_name | any
property        ::= property_simple_name [(value {, value})]
integer_exp     ::= integer_expression
                    | integer_placeholder | integer_property
                    | ( integer_exp )
                    | integer_exp binary_op integer_exp
                    | unary_op integer_exp
discrete_exp    ::= discrete_expression | discrete_property
                    | ( discrete_exp ) | integer_exp | boolean_exp
```

```
boolean_exp           ::= boolean_expression | boolean_property
                        | ( boolean_exp )
                        | task_name equal_op task_name
                        | entry_name equal_op entry_name
                        | action_name equal_op action_name
                        | task_type_name equal_op task_type_name
                        | discrete_exp relation discrete_exp
                        | not boolean_exp
                        | boolean_exp logical_op boolean_exp
binary_op             ::= + | - | * | / | mod | rem | **
unary_op              ::= + | - | abs
equal_op              ::= = | /=
relation_op           ::= equal_op | > | < | <= | >=
logical_op            ::= and | or | xor
tsl_type              ::= task | entry | task_type | action
                        | task_ada_type | discrete_type
```

Events

```
guard                 ::= where boolean_exp
guarded_event         ::= basic_event [guard] | timer
basic_event           ::= call | accept | start_rendezvous
                          | end_rendezvous | termination
                          | activation | performs
call                  ::= task_name calls task_name
                              [at entry_name [binding]]
accept                ::= task_name accepting
                              [entry_name {, entry_name}]
start_rendezvous      ::= task_name accepts [task_name]
                              [at entry_name [binding]]
end_rendezvous        ::= task_name releases [task_name]
                              [from entry_name [binding]]
termination           ::= task_name terminates
activation            ::= task_name activates task_name
                              [of task_type_name]
performs              ::= task_name performs action_simple_name
                              [binding]
                          | task_name performs action_name
timer                 ::= integer_exp milliseconds
binding               ::= (simple_name => tsl_exp
                              {,simple_name => tsl_exp})
```

Compound Events

```
compound_event  ::= guarded_event
                  | (compound_event)
                  | macro_call
                  | (compound_event) ↑ integer_exp
                  | compound_event => compound_event
                  | compound_event and compound_event
                  | compound_event or compound_event
macro_call      ::= macro_simple_name [(value {, value})]
```

Patterns

```
specification ::= [<<simple_name>>] [activator] [not]
                    compound_event [terminator];
update        ::= [<<simple_name>>] [activator]
                  [compound_event]
                    set property := value {, property := value};
activator     ::= when compound_event then
terminator    ::= before compound_event
                  | until compound_event
```

Definitions

```
macro_def    ::= macro simple_name
                 [(placeholder {, placeholder} : tsl_type]
                 {; placeholder {, placeholder} : tsl_type})]
                  is compound_event ;
action_def   ::= action simple_name
                 [(simple_name {simple_name} : tsl_type
                 {; simple_name {, simple_name} : tsl_type})];
property_def ::= property simple_name
                 [(tsl_type {, tsl_type} )] : tsl_type
                 := value [is update {update} end] ;
```

Declarations

```
declaration ::= specification | macro_def | action_def
                  | property_def
```

Statements

```
event_generator ::= perform action_simple_name
                          [(value {, value})];
```

TSL Reserved Words

The following is the list of TSL reserved words which are not also reserved words of Ada.

accepting	accepts	action	activates	any
before	calls	from	macro	main
milliseconds	perform	performs	property	releases
self	set	terminates	until	where

Predefined Environment

```
property Running (task)        : Boolean := False;
property Calling (task)        : Boolean := False;
property In_Rendezvous (task) : Boolean := False;
property Blocked (task)        : Boolean := False;
property Terminated (task)    : Boolean := False;
property Is_Accepting (task) : Boolean := False;
property Open_Entry (task, entry) : Boolean := False;
property Task_Called (task)  : task := null;
property Entry_Called (task) : entry := null;
property Queue_Size (task, entry) : Integer := 0;
property Type_Of (task)       : task_type := null;

action Calls (The_Task : task; At_Entry : entry);
action Accepting (Entry_1, Entry_2, Entry_3, ... : entry);
action Accepts   (The_Task : task; At_Entry : entry);
action Releases  (The_Task : task; From_Entry : entry);
action Terminates;
action Activates (The_Task : task; Of_Type : task_type);
```

References

[1] *The Ada Programming Language Reference Manual*
     US Department of Defense, US Government Printing Office, 1983.
     ANSI/MILSTD 1815A Document.

[2] Bates, P.C. and Wileden, J.C.
     EDL: A Basis for Distributed System Debugging Tools.
     In *Proceedings of Hawaii International Conference on System Sciences*,
         pages 86-93. Hawaii International Conference on System Sciences,
         Honolulu, Hawaii, January, 1982.

[3] Campbell,R.H. and Habermann, A.N.
     The Specification of Process Synchronization by Path-Expressions.
     Lecture Notes in Computer Science 16, 1974.

[4] Clocksin, W.F., and Mellish, C.S.
     Programming in Prolog.
     Springer-Velag, 1981.

[5] Helmbold, D.P. and Luckham, D.C.
     Debugging Ada Tasking Programs.
     In *Proceedings of the IEEE Computer Society 1984 Conference on Ada
         Applications and Environments*, pages 96-110. IEEE, St. Paul,
         Minnesota, October 15-18, 1984.
     Also published, Stanford University Computer Systems Laboratory TR
         84-262, July, 1984, Program Analysis and Verification Group Report
         25.

[6] Helmbold, D.P., and Luckham, D.C.
     Runtime Detection and Description of Deadness Errors in Ada Tasking.
     CSL Technical Report 83-249, Stanford University, November, 1983.
     Program Analysis and Verification Group Report 22.

[7] Helmbold, D.P., and Luckham, D.C.
     Debugging Ada Tasking Programs.
     IEEE Software 2(2):47-57, March, 1985.
     In Proceedings of the IEEE Computer Society 1984 Conference on Ada
         Applications and Environments, pp.96-110. IEEE, St. Paul,
         Minnesota, October 15-18, 1984. Also published as Stanford University
         CSL TR.84-263, July, 1984.

[8]   Helmbold, D.P., and Luckham, D.C.
      TSL: Task Sequencing Language.
      In *Proceedings of the 1985 SIGAda International Conference*, pages
          255-274. ACM, Paris, France, May, 1985.
      Also published in a special edition of *Ada Letters*, Vol.V, Issue 2,
          September-October 1985.

[9]   Hewitt, C.
      *Planner: A Language for Proving Theorems and Manipulating Models in
          a Robot.*
      PhD thesis, Massachusetts Institute of Technology, January, 1971.

[10] Kahn, G. and MacQueen, D.
      Coroutines and Networks of Parallel Processes.
      In *Proceedings of IFIP Congress '77*, pages 993-998. North-Holland
          Publishing Company, Amsterdam, August, 1977.

[11] Lamport, L., Shostak, R., and Pease, M.
      The Byzantine Generals Problem.
      Transactions of Programming Languages and Systems 4(3):382-401, July,
          1982.

[12] Ledoux, C., and Parker, D.S.
      Saving Traces for Ada Debugging.
      In *Proceedings of the Ada International Conference'85*, pages 97-108.
          Cambridge University Press, Month, 1985.

[13] Dahl, O.-J.
      Time Sequences as a Tool For Describing Program Behaviour.
      Research Report in Informatics 48, University of Oslo, August, 1979.

[14] Peterson, J.L.
      Petri Nets.
      Computing Surveys 9(3), September, 1977.

Vol. 245: H.F. de Groote, Lectures on the Complexity of Bilinear Problems. V, 135 pages. 1987.

Vol. 246: Graph-Theoretic Concepts in Computer Science. Proceedings, 1986. Edited by G. Tinhofer and G. Schmidt. VII, 307 pages. 1987.

Vol. 247: STACS 87. Proceedings, 1987. Edited by F.J. Brandenburg, G. Vidal-Naquet and M. Wirsing. X, 484 pages. 1987.

Vol. 248: Networking in Open Systems. Proceedings, 1986. Edited by G. Müller and R.P. Blanc. VI, 441 pages. 1987.

Vol. 249: TAPSOFT '87. Volume 1. Proceedings, 1987. Edited by H. Ehrig, R. Kowalski, G. Levi and U. Montanari. XIV, 289 pages. 1987.

Vol. 250: TAPSOFT '87. Volume 2. Proceedings, 1987. Edited by H. Ehrig, R. Kowalski, G. Levi and U. Montanari. XIV, 336 pages. 1987.

Vol. 251: V. Akman, Unobstructed Shortest Paths in Polyhedral Environments. VII, 103 pages. 1987.

Vol. 252: VDM '87. VDM – A Formal Method at Work. Proceedings, 1987. Edited by D. Bjørner, C.B. Jones, M. Mac an Airchinnigh and E.J. Neuhold. IX, 422 pages. 1987.

Vol. 253: J.D. Becker, I. Eisele (Eds.), WOOPLOT 86. Parallel Processing: Logic, Organization, and Technology. Proceedings, 1986. V, 226 pages. 1987.

Vol. 254: Petri Nets: Central Models and Their Properties. Advances in Petri Nets 1986, Part I. Proceedings, 1986. Edited by W. Brauer, W. Reisig and G. Rozenberg. X, 480 pages. 1987.

Vol. 255: Petri Nets: Applications and Relationships to Other Models of Concurrency. Advances in Petri Nets 1986, Part II. Proceedings, 1986. Edited by W. Brauer, W. Reisig and G. Rozenberg. X, 516 pages. 1987.

Vol. 256: Rewriting Techniques and Applications. Proceedings, 1987. Edited by P. Lescanne. VI, 285 pages. 1987.

Vol. 257: Database Machine Performance: Modeling Methodologies and Evaluation Strategies. Edited by F. Cesarini and S. Salza. X, 250 pages. 1987.

Vol. 258: PARLE, Parallel Architectures and Languages Europe. Volume I. Proceedings, 1987. Edited by J.W. de Bakker, A.J. Nijman and P.C. Treleaven. XII, 480 pages. 1987.

Vol. 259: PARLE, Parallel Architectures and Languages Europe. Volume II. Proceedings, 1987. Edited by J.W. de Bakker, A.J. Nijman and P.C. Treleaven. XII, 464 pages. 1987.

Vol. 260: D.C. Luckham, F.W. von Henke, B. Krieg-Brückner, O. Owe, ANNA, A Language for Annotating Ada Programs. V, 143 pages. 1987.

Vol. 261: J. Ch. Freytag, Translating Relational Queries into Iterative Programs. XI, 131 pages. 1987.

Vol. 262: A. Burns, A.M. Lister, A.J. Wellings, A Review of Ada Tasking. VIII, 141 pages. 1987.

Vol. 263: A.M. Odlyzko (Ed.), Advances in Cryptology – CRYPTO '86. Proceedings. XI, 489 pages. 1987.

Vol. 264: E. Wada (Ed.), Logic Programming '86. Proceedings, 1986. VI, 179 pages. 1987.

Vol. 265: K.P. Jantke (Ed.), Analogical and Inductive Inference. Proceedings, 1986. VI, 227 pages. 1987.

Vol. 266: G. Rozenberg (Ed.), Advances in Petri Nets 1987. VI, 451 pages. 1987.

Vol. 267: Th. Ottmann (Ed.), Automata, Languages and Programming. Proceedings, 1987. X, 565 pages. 1987.

Vol. 268: P.M. Pardalos, J.B. Rosen, Constrained Global Optimization: Algorithms and Applications. VII, 143 pages. 1987.

Vol. 269: A. Albrecht, H. Jung, K. Mehlhorn (Eds.), Parallel Algorithms and Architectures. Proceedings, 1987. Approx. 205 pages. 1987.

Vol. 270: E. Börger (Ed.), Computation Theory and Logic. IX, 442 pages. 1987.

Vol. 271: D. Snyers, A. Thayse, From Logic Design to Logic Programming. IV, 125 pages. 1987.

Vol. 272: P. Treleaven, M. Vanneschi (Eds.), Future Parallel Computers. Proceedings, 1986. V, 492 pages. 1987.

Vol. 273: J.S. Royer, A Connotational Theory of Program Structure. V, 186 pages. 1987.

Vol. 274: G. Kahn (Ed.), Functional Programming Languages and Computer Architecture. Proceedings. VI, 470 pages. 1987.

Vol. 275: A.N. Habermann, U. Montanari (Eds.), System Development and Ada. Proceedings, 1986. V, 305 pages. 1987.